Recent Titles in This Series

(Continued in the back of this publication)

WITHDRAWN

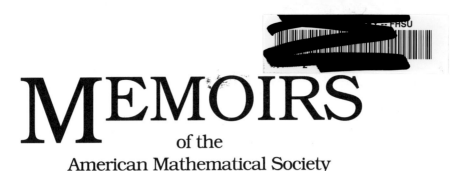

MEMOIRS
of the
American Mathematical Society

Number 516

A Proof of the q-Macdonald-Morris Conjecture for BC_n

Kevin W. J. Kadell

March 1994 • Volume 108 • Number 516 (first of 5 numbers) • ISSN 0065-9266

American Mathematical Society
Providence, Rhode Island

1991 *Mathematics Subject Classification.*
Primary 17B20.

Library of Congress Cataloging-in-Publication Data

Kadell, Kevin W. J., 1950–
  A proof of the q-Macdonald-Morris conjecture for BC_n/Kevin W. J. Kadell.
    p. cm. – (Memoirs of the American Mathematical Society; no. 516)
  Includes bibliographical references.
  ISBN 0-8218-2552-6
  1. Beta functions. 2. Integrals, Definite. 3. Selberg trace formula. I. Title. II. Series.
QA3.A57 no. 516
[QA351]
510 s–dc20   93-48293
[515′.52]   CIP

Memoirs of the American Mathematical Society

This journal is devoted entirely to research in pure and applied mathematics.

Subscription information. The 1994 subscription begins with Number 512 and consists of six mailings, each containing one or more numbers. Subscription prices for 1994 are $353 list, $282 institutional member. A late charge of 10% of the subscription price will be imposed on orders received from nonmembers after January 1 of the subscription year. Subscribers outside the United States and India must pay a postage surcharge of $25; subscribers in India must pay a postage surcharge of $43. Expedited delivery to destinations in North America $30; elsewhere $92. Each number may be ordered separately; *please specify number* when ordering an individual number. For prices and titles of recently released numbers, see the New Publications sections of the *Notices of the American Mathematical Society.*

Back number information. For back issues see the *AMS Catalog of Publications.*

Subscriptions and orders should be addressed to the American Mathematical Society, P. O. Box 5904, Boston, MA 02206-5904. *All orders must be accompanied by payment.* Other correspondence should be addressed to Box 6248, Providence, RI 02940-6248.

Copying and reprinting. Individual readers of this publication, and nonprofit libraries acting for them, are permitted to make fair use of the material, such as to copy a chapter for use in teaching or research. Permission is granted to quote brief passages from this publication in reviews, provided the customary acknowledgement of the source is given.

Republication, systematic copying, or multiple reproduction of any material in this publication (including abstracts) is permitted only under license from the American Mathematical Society. Requests for such permission should be addressed to the Manager of Editorial Services, American Mathematical Society, P. O. Box 6248, Providence, RI 02940-6248. Requests can also be made by e-mail to `reprint-permission@math.ams.org`.

The owner consents to copying beyond that permitted by Sections 107 or 108 of the U.S. Copyright Law, provided that a fee of $1.00 plus $.25 per page for each copy be paid directly to the Copyright Clearance Center, Inc., 222 Rosewood Dr., Danvers, MA 01923. When paying this fee please use the code 0065-9266/94 to refer to this publication. This consent does not extend to other kinds of copying, such as copying for general distribution, for advertising or promotion purposes, for creating new collective works, or for resale.

Memoirs of the American Mathematical Society is published bimonthly (each volume consisting usually of more than one number) by the American Mathematical Society at 201 Charles Street, Providence, RI 02904-2213. Second-class postage paid at Providence, Rhode Island. Postmaster: Send address changes to Memoirs, American Mathematical Society, P. O. Box 6248, Providence, RI 02940-6248.

ⓒ Copyright 1994, American Mathematical Society. All rights reserved.
Printed in the United States of America.
This volume was printed directly from author-prepared copy.
♾ The paper used in this book is acid-free and falls within the guidelines
established to ensure permanence and durability.
♻ Printed on recycled paper.

10 9 8 7 6 5 4 3 2 1   99 98 97 96 95 94

TABLE OF CONTENTS

ABSTRACT

Macdonald and Morris gave a series of constant term q-conjectures associated with root systems. Selberg evaluated a multivariable beta type integral which plays an important role in the theory of constant term identities associated with root systems. Aomoto recently gave a simple and elegant proof of a generalization of Selberg's integral. Kadell extended this proof to treat Askey's conjectured q-Selberg integral, which was proved independently by Habsieger. We use a constant term formulation of Aomoto's argument to treat the q-Macdonald-Morris conjecture for the root system BC_n. The proof is based upon the fact that if $f(t_1, \ldots, t_n)$ has a Laurent expansion at $t_1 = 0$, then the constant term of $f(t_1, \ldots, t_n)$ is fixed by $t_1 \rightarrow qt_1$. The q-engine of our q-machine is the equivalent conclusion that $\partial_q/\partial_q t_1 f(t_1, \ldots, t_n)$ has no residue at $t_1 = 0$. We use an identity for a partial q-derivative which owes its existence to the geometry of the simple roots of B_n and C_n. We also require certain antisymmetries of the terms occurring in the partial q-derivative and the q-transportation theory for BC_n. These are proved locally by using the basic property of the simple reflections of B_n and C_n that the reflection along a simple positive root $\alpha \in R^+$ sends α to $-\alpha$ and permutes the other positive roots. We show how to obtain the required functional equations using only the q-transportation theory for BC_n. This is based upon the fact that B_n and C_n have the same Weyl group. The B_n, $B_n^\vee$ and D_n cases of the conjecture follow from our theorem for BC_n. We give some of the details for C_n and $C_n^\vee$ and acknowledge the priority of Gustafson's recent proof of these cases of the conjecture. This illustrates the basic steps required to apply our methods to the conjecture when the reduced irreducible root system R does not have a miniscule weight.

Much of the research for this paper was done in 1986 while on a visiting appointment to the Department of Combinatorics and Optimization, University of Waterloo, Waterloo, Ontario, Canada N2L 3G1. Research supported by NSERC grants A8907 and A8235 and National Science Foundation grant DMS-87-01967.

Received by the editor May 5, 1989 and in revised form November 6, 1991.

1. Introduction

Macdonald [Ma1] and Morris [Mo1] gave a series of constant term q-conjectures associated with root systems. Macdonald expressed these conjectures in terms of the algebra of root systems, while Morris stated many of them explicitly. In 1944, Selberg [Se1] evaluated a multivariable beta type integral which plays an important role in the theory of constant term identities associated with root systems. Aomoto [Ao1] recently gave a simple and elegant proof of a generalization of Selberg's integral. Kadell [Ka3] extended this proof to treat Askey's [As1] conjectured q-Selberg integral, which was proved independently by Habsieger [Ha2]. We use a constant term formulation of Aomoto's argument to treat the q-Macdonald-Morris conjecture for BC_n. The proof is based upon the fact that if $f(t_1, \ldots, t_n)$ has a Laurent expansion at $t_1 = 0$, then the constant term of $f(t_1, \ldots, t_n)$ is fixed by $t_1 \to qt_1$. The q-engine of our q-machine is the equivalent conclusion that $\partial_q/\partial_q t_1 f(t_1, \ldots, t_n)$ has no residue at $t_1 = 0$. We use an identity for a partial q-derivative which owes its existence to the geometry of the simple roots of B_n and C_n. We also require certain antisymmetries of the terms occurring in the partial q-derivative and the q-transportation theory for BC_n. These are proved locally by using the basic property of the simple reflections of B_n and C_n that the reflection along a simple positive root $\alpha \in R^+$ sends α to $-\alpha$ and permutes the other positive roots. We show how to obtain the required functional equations using only the q-transportation theory for BC_n. This is based upon the fact that B_n and C_n have the same Weyl group.

The B_n, $B_n^{\vee}$ and D_n cases of the q-Macdonald-Morris conjecture follow from our theorem for BC_n. We give some of the details for C_n and $C_n^{\vee}$ and acknowledge the priority of Gustafson's recent proof of these cases of the conjecture. This illustrates the basic steps required to apply our methods to the conjecture when the reduced irreducible root system R does not have a minuscule weight.

Let V be a vector space over the reals with dimension ℓ, positive definite symmetric bilinear form (α, β), and orthonormal basis $\{e_1, \ldots, e_\ell\}$. Let R be an irreducible, not necessarily reduced, root system on V and let R^+ be a system of positive roots. Let σ_α denote the reflection along $\alpha \in R$ and let W be the Weyl group of R. Let e^α denote the formal exponential of $\alpha \in R$ and let k_α be a nonnegative integer for all $\alpha \in R$ such that

$$(1.1) \qquad k_\alpha = k_\beta \text{ if } |\alpha| = |\beta|.$$

This is natural since W acts transitively on roots of a given length.

Macdonald [Ma1] conjectured a constant term identity associated with the root system R. It was recently proved by Opdam [Op3] (see also [Op1, Op2], Heckman [He1], and Heckman and Opdam [HO1]).

Theorem 1. (Opdam [Op3])

$$(1.2) \qquad [1] \prod_{\alpha \in R} (1 - e^\alpha)^{k_\alpha} = \prod_{\alpha \in R} \frac{(|(\rho_k, \alpha^{\vee}) + k_\alpha + \frac{1}{2} k_{\alpha/2}|)!}{(|(\rho_k, \alpha^{\vee}) + \frac{1}{2} k_{\alpha/2}|)!},$$

where $[w]f$ is the coefficient of the monomial w in the Laurent expansion of f,

$$(1.3) \qquad \rho_k = \frac{1}{2} \sum_{\alpha \in R^+} k_\alpha \alpha,$$

1

$\alpha^{\vee} = 2\alpha/|\alpha|^2$ is the coroot corresponding to α, and $k_{\alpha/2} = 0$ if $\alpha/2 \notin R$.

For the infinite families of irreducible root systems, we have the systems

(1.4)
$$
\begin{aligned}
A_{n-1}^+ &= \{e_i - e_j \mid 1 \le i < j \le n\}, \\
D_n^+ &= \{e_i + e_j \mid 1 \le i < j \le n\} \bigcup A_{n-1}^+, \\
B_n^+ &= \{e_i \mid 1 \le i \le n\} \bigcup D_n^+, \\
C_n^+ &= \{2e_i \mid 1 \le i \le n\} \bigcup D_n^+, \\
BC_n^+ &= B_n^+ \bigcup C_n^+,
\end{aligned}
$$

of positive roots.

We take the standard formal exponential $e^{e_i} = t_i$, $1 \le i \le n$. The A_{n-1} case of Theorem 1 is

(1.5)
$$
[1] \prod_{1 \le i < j \le n} (1 - \frac{t_i}{t_j})^k (1 - \frac{t_j}{t_i})^k = \frac{(nk)!}{(k!)^n},
$$

while the BC_n case is

(1.6)
$$
\begin{aligned}
& [1] \prod_{i=1}^n (1 - t_i)^a (1 - \frac{1}{t_i})^a (1 - t_i^2)^b (1 - \frac{1}{t_i^2})^b \\
& \times \prod_{1 \le i < j \le n} (1 - \frac{t_i}{t_j})^k (1 - \frac{t_j}{t_i})^k (1 - t_i t_j)^k (1 - \frac{1}{t_i t_j})^k \\
&= \prod_{i=1}^n \frac{(2b + 2(n-i)k)! \, (2a + 2b + 2(n-i)k)!}{(b + (n-i)k)! \, (a + b + (n-i)k)! \, (a + 2b + (2n-i-1)k)!} \frac{(ik)!}{k!}.
\end{aligned}
$$

The B_n, C_n and D_n cases are obtained by setting $b = 0$, $a = 0$ and $a = b = 0$, respectively, in the BC_n case (1.6).

Let $n \ge 1$ be a positive integer, $\operatorname{Re}(x) > 0$, $\operatorname{Re}(y) > 0$, and let m and ℓ be nonnegative integers satisfying $m + \ell \le n$. One may add the parameter ℓ to Aomoto's extension [Ao1] of Selberg's integral [Se1]. This is given by the following theorem.

Theorem 2. (Aomoto [Ao1])
(1.7)
$$
\begin{aligned}
I_{n,m,\ell}^k & (x, y) \\
&= \int_0^1 \cdots \int_0^1 \prod_{i=1}^n t_i^{(x-1)+\chi(i \le m)} (1 - t_i)^{(y-1)+\chi(n-i+1 \le \ell)} \Delta_n^{2k}(t_1, \dots, t_n) \, dt_1 \dots dt_n \\
&= \prod_{i=1}^n \frac{\Gamma(x + (n-i)k + \chi(i \le m)) \, \Gamma(y + (n-i)k + \chi(i \le \ell)) \, \Gamma(1 + ik)}{\Gamma(x + y + (2n-i-1)k + \chi(i \le m + \ell))} \frac{\Gamma(1 + ik)}{\Gamma(1 + k)},
\end{aligned}
$$

where $\chi(A)$ is 1 or 0 according to whether A is true or false, respectively, and

$$(1.8) \qquad \Delta_n(t_1, \ldots, t_n) = \prod_{1 \le i < j \le n} (t_i - t_j)$$

denotes the Vandermonde determinant.

We omit ℓ when $\ell = 0$ and we omit m and ℓ when $m = \ell = 0$.

The substitution $t_i \to (1 - t_{n-i+1})$, $1 \le i \le n$, gives the symmetry

$$(1.9) \qquad I_{n,m,\ell}^k(x,y) = I_{n,\ell,m}^k(y,x).$$

This is essential to Selberg's proof [Se1] of his integral, which is the $m = \ell = 0$ case of (1.7). See Andrews [An2] for a readily accessible version of this proof. Aomoto's simple and elegant proof [Ao1] uses a functional equation involving the parameter m. This requires the symmetry in $t_1, \ldots, t_n$ of the integrand of Selberg's integral.

Using an idea of Regev, Macdonald [Ma1] showed that the BC_n case (1.6) of Theorem 1 is equivalent to Selberg's integral. Morris [Mo1, (4.13)] proved an extension of (1.5) which is associated with the root system A_n and is also equivalent to Selberg's integral. It is

$$(1.10) \qquad [1] \prod_{i=1}^{n} (1 - t_i)^a (1 - \frac{1}{t_i})^b \prod_{1 \le i < j \le n} (1 - \frac{t_i}{t_j})^k (1 - \frac{t_j}{t_i})^k$$
$$= \prod_{i=1}^{n} \frac{(a + b + (n-i)k)!}{(a + (n-i)k)!\,(b + (n-i)k)!} \frac{(ik)!}{k!}.$$

Morris [Mo1] independently discovered the role (1.1) of root lengths and the connection between Selberg's integral and BC_n. He used MACSYMA to obtain explicit q-analogues of Theorem 1 for BC_n [Mo1, (3.4)] and G_2 [Mo1, Appendix C]. Macdonald incorporated these into a general q-conjecture [Ma1, Conjecture 3.1''], also given by Morris [Mo1, (3.6)], associated with the affine root system $S(R)$. We call [Mo1, Conjecture A'] the q-Macdonald-Morris conjecture for R. Macdonald [Ma1] proved the conjecture when R is reduced and k_α identically equals 1, 2 or ∞. Many cases of this conjecture are given explicitly by Morris [Mo1].

Let q be fixed with $0 < q < 1$ and

$$(x)_0 = (x;q)_0 = 1,$$

$$(1.11) \qquad (x)_n = (x;q)_n = \prod_{i=0}^{n-1} (1 - xq^i), \quad n \ge 1,$$

$$(x)_\infty = (x;q)_\infty = \lim_{n \to \infty} (x;q)_n = \prod_{i \ge 0} (1 - xq^i).$$

For the q-Macdonald-Morris conjecture, the product $\prod_{\alpha \in R} (1 - e^\alpha)^{k_\alpha}$ on the left side of (1.2) becomes

$$(1.12) \qquad {}_q r_\ell(k; e^{e_1}, \ldots, e^{e_\ell}) = \prod_{\alpha \in R^+} (q^{m_\alpha} e^\alpha; q^{u_\alpha})_{k_\alpha} (q^{u_\alpha - m_\alpha} e^{-\alpha}; q^{u_\alpha})_{k_\alpha},$$

where m_α and u_α are associated with the affine root system $S(R)$ and depend only on the length of α. Observe that the positive and negative roots are treated differently in $_q r_\ell(k; e^{e_1}, \ldots, e^{e_\ell})$. The q-Macdonald-Morris conjecture involves the constant term

$$(1.13) \qquad _q R_\ell(k) = [1]\, _q r_\ell(k; e^{e_1}, \ldots, e^{e_\ell}).$$

Independently, Kadell [Ka3] and Habsieger [Ha2] proved a q-analogue of Selberg's integral conjectured by Askey [As1]. Kadell used Aomoto's argument [Ao1] to obtain a q-analogue of (1.7). Habsieger extended Selberg's proof [Se1] by using a clever asymptotic argument in place of the symmetry (1.9). They observed that their result is equivalent to a constant term identity conjectured by Morris [Mo1, (4.12)], which is a q-analogue of (1.10). This is given by the following theorem.

Theorem 3. (Habsieger [Ha2], Kadell [Ka3])

$$(1.14) \quad [1]\prod_{i=1}^{n}(t_i)_a\left(\frac{q}{t_i}\right)_b \prod_{1\le i<j\le n}\left(\frac{t_i}{t_j}\right)_k\left(q\frac{t_j}{t_i}\right)_k = \prod_{i=1}^{n}\frac{(q)_{a+b+(n-i)k}}{(q)_{a+(n-i)k}\,(q)_{b+(n-i)k}}\frac{(q)_{ik}}{(q)_k}.$$

See Stembridge [St1] and Zeilberger [Ze3] for additional proofs.

Let us survey the known results for the isolated root systems. Zeilberger [Ze1] and Habsieger [Ha1] independently showed that the q-Macdonald-Morris conjecture for G_2 follows from the $n = 3$ case of the q-Selberg integral. Zeilberger used the q-Morris theorem (1.14). Garvan [Ga1] gave an Aomoto-type extension for G_2 when $q = 1$ which had been conjectured by Askey [As2]. Garvan [Ga2] proved the $q = 1$ case of the conjecture for F_4. Garvan and Gonnet [GG1] proved the conjecture for F_4 and $F_4^\vee$. Zeilberger [Ze2] proved the conjecture for $G_2^\vee$. Hence the q-Macdonald-Morris conjecture remains open for the isolated root systems E_6, E_7 and E_8.

For the root system BC_n we have

$$(1.15) \qquad \begin{aligned} m_\alpha &= 0,\ u_\alpha = 1,\ \text{if } |\alpha|^2 = 1 \text{ or } 2,\\ m_\alpha &= 1,\ u_\alpha = 2,\ \text{if } |\alpha|^2 = 4. \end{aligned}$$

Let $n \ge 1$ and let k, a, b, m and r be nonnegative integers with $m+r \le n$. Following Morris [Mo1, (3.4)], we set

$$(1.16)$$
$$_q bc_{n,m,r}(a,b,k;t_1,\ldots,t_n) = \prod_{i=1}^{n}t_i^{\chi(i\le r)}(t_i)_{a+\chi(n-i+1\le m)}\left(\frac{q}{t_i}\right)_a(qt_i^2;q^2)_b\left(\frac{q}{t_i^2};q^2\right)_b$$
$$\times \prod_{1\le i<j\le n}\left(\frac{t_i}{t_j}\right)_k\left(q\frac{t_j}{t_i}\right)_k(t_it_j)_k\left(\frac{q}{t_it_j}\right)_k$$

and we denote the constant term by

$$(1.17) \qquad _q BC_{n,m,r}(a,b,k) = [1]\, _q bc_{n,m,r}(a,b,k;t_1,\ldots,t_n).$$

We omit r when $r = 0$ and we omit m and r when $m = r = 0$. We also omit q when $q = 1$.

The q-Macdonald-Morris conjecture for BC_n is the $m = 0$ case of the following theorem, which is our main result.

Theorem 4.

(1.18)

$$_qBC_{n,m}(a,b,k) = \prod_{i=1}^{n} \frac{(q)_{2a+2b+2(n-i)k+\chi(i\leq m)}}{(q)_{a+2b+(2n-i-1)k+\chi(i\leq m)}} \frac{(q)_{2b+2(n-i)k} \ (q)_{ik}}{(q)_{a+(n-i)k} \ (q)_{2b+(n-i)k} \ (q)_k}$$

$$\times \prod_{i=1}^{n} \frac{(q^2;q^2)_{2b+(n-i)k} \ (q^2;q^2)_{a+(n-i)k}}{(q^2;q^2)_{b+(n-i)k} \ (q^2;q^2)_{a+b+(n-i)k}}.$$

Given the central role that Selberg's integral plays in constant term identities associated with root systems, Zeilberger suggested in a post card that the q-version [Ka3] of Aomoto's proof [Ao1] might be modified to treat the q-Macdonald-Morris conjecture for BC_n. That is, the $m = 0$ case of Theorem 4 (1.18). This indeed happens and the parameter m in Theorem 4 (1.18) does correspond to the parameter m in Aomoto's Theorem 2 (1.7).

Dyson marvelled in a letter to Morris at the discovery of the wondrous function $_qbc_n(a,b,k;t_1,\ldots,t_n)$ and its constant term $_qBC_n(a,b,k)$. The elegant creation of Morris $_qbc_n(a,b,k;t_1,\ldots,t_n)$ incorporates the role that simple roots and reflections play in the relationships between the root systems A_{n-1}, D_n, B_n, C_n and BC_n. We explicitly express the geometry of the simple roots and the basic property of the simple reflections of B_n and C_n in terms of $_qbc_n(a,b,k;t_1,\ldots,t_n)$. We show that $_qbc_n(a,b,k;t_1,\ldots,t_n)$ satisfies certain near symmetries which are just what we require to do the analysis of the proof of Theorem 4 locally.

We use a constant term formulation of Aomoto's argument [Ao1] to treat the $q = 1$ case of Theorem 4. The engine of our machine is the fact that if $f(t_1,\ldots,t_n)$ has a Laurent expansion at $t_1 = 0$, then $\partial/\partial t_1 f(t_1,\ldots,t_n)$ has no residue at $t_1 = 0$. We require the fact that $bc_n(a,b,k;t_1,\ldots,t_n)$ is symmetric in $t_1,\ldots,t_n$ and invariant under $t_i \leftrightarrow 1/t_i$ for $1 \leq i \leq n$.

Our proof of Theorem 4 is based upon the fact that if $f(t_1,\ldots,t_n)$ has a Laurent expansion at $t_1 = 0$, then the constant term $[1]f(t_1,\ldots,t_n)$ of $f(t_1,\ldots,t_n)$ is fixed by $t_1 \to qt_1$. The q-engine of our q-machine is the equivalent conclusion that $\partial_q/\partial_q t_1 f(t_1,\ldots,t_n)$ has no residue at $t_1 = 0$. Since the positive and negative roots are treated differently, $_qbc_n(a,b,k;t_1,\ldots,t_n)$ is not symmetric in $t_1,\ldots,t_n$ or invariant under $t_i \to 1/t_i$ for $1 \leq i \leq n$. However, it satisfies certain near symmetries which are just what we require to do the analysis of the proof of Theorem 4 locally.

We use an identity for a partial q-derivative which owes its existence to the geometry of the simple roots of B_n and C_n. We also require certain antisymmetries of the factors arising in the partial q-derivative and the q-transportation theory for BC_n. These are proved locally by using the basic property of the simple reflections of B_n and C_n that the reflection along a simple positive root $\alpha \in R^+$ sends α to $-\alpha$ and permutes the other positive roots. We show how to obtain the required functional equations using only the q-transportation theory for BC_n. This is based upon the fact that B_n and C_n have the same Weyl group.

The B_n, $B_n^\vee$ and D_n cases of the q-Macdonald-Morris conjecture follow from our theorem for BC_n. We give some of the details for C_n and $C_n^\vee$ and acknowledge the priority of Gustafson's recent proof of these cases of the conjecture. This illustrates the basic steps required to apply our methods to the conjecture when the reduced irreducible root system R does not have a minuscule weight.

2. Outline of the proof and summary

In this section, we use a constant term formulation of Aomoto's argument [Ao1] to treat the $q = 1$ case of Theorem 4 and summarize the results of each section of the paper.

The engine of Aomoto's proof [Ao1] of his extension Theorem 2 (1.7) of Selberg's integral is the fact that if

$$(2.1) \qquad F(0, t_2, \dots, t_n) = F(1, t_2, \dots, t_n)$$

and $F(t_1, \dots, t_n)$ satisfies some simple conditions, then applying the fundamental theorem of calculus to the variable t_1 gives

$$(2.2) \qquad \int_0^1 \cdots \int_0^1 \frac{\partial}{\partial t_1} F(t_1, \dots, t_n) \, dt_1 \dots dt_n = 0.$$

The integrand

$$(2.3) \qquad W_n^k(x, y; t_1, \dots, t_n) = \prod_{i=1}^n t_i^{(x-1)} (1 - t_i)^{(y-1)} \Delta_n^{2k}(t_1, \dots, t_n)$$

of Selberg's integral is a symmetric function of $t_1, \dots, t_n$. That is

$$(2.4) \qquad W_n^k(x, y; t_{\pi(1)}, \dots, t_{\pi(n)}) = W_n^k(x, y; t_1, \dots, t_n), \quad \pi \in S_n,$$

where S_n is the symmetric group on n letters. Thus we have

$$(2.5) \qquad
\begin{aligned}
I_{n,m}^k(x, y) &= \int_0^1 \cdots \int_0^1 \prod_{i=1}^m t_i \, W_n^k(x, y; t_1, \dots, t_n) \, dt_1 \dots dt_n \\
&= \int_0^1 \cdots \int_0^1 \prod_{i=1}^m t_{\pi(i)} \, W_n^k(x, y; t_1, \dots, t_n) \, dt_1 \dots dt_n.
\end{aligned}$$

We let $\pi \in S_n$ correspond to the substitution $t_i \to t_{\pi(i)}$, $1 \le i \le n$. Since the Weyl group of the root system A_{n-1} is S_n, we call (2.5) the integral formulation of the transportation property for A_{n-1}.

Aomoto [Ao1] applies the engine (2.2) to $f(t_1, \dots, t_n) = (1 - t_1) \prod_{i=1}^m t_i \, W_n^k(x, y; t_1, \dots, t_n)$, $m \ge 1$. Using the symmetry (2.4) of $W_n^k(x, y; t_1, \dots, t_n)$ and the transportation property (2.5), he obtains the functional equation

$$(2.6) \qquad I_{n,m}^k(x, y) = \frac{(x + (n - m)k)}{(x + y + (2n - m - 1)k)} \, I_{n,m-1}^k(x, y), \quad m \ge 1.$$

Since

$$(2.7) \qquad I_{n,n}^k(x, y) = I_n^k(x + 1, y),$$

we see that the parameter m refines the dependence of Selberg's integral on x. The $\ell = 0$ case of Theorem 2 follows using the recurrence relation

$$(2.8) \qquad \lim_{x \to 0} x\, I_n^k(x, y) = n\, I_{n-1}^k(2k, y)$$

and a double induction on x and n. Since $1 = t_i + (1 - t_i)$, we have

$$(2.9) \qquad I_{n,m,\ell}^k(x, y) = I_{n,m+1,\ell}^k(x, y) + I_{n,m,\ell+1}^k(x, y), \quad m + \ell + 1 \le n.$$

Theorem 2 now follows by induction on ℓ. See Askey [As2] for more of the details of Aomoto's proof.

We may reformulate Aomoto's proof [Ao1] to treat constant term identities by replacing the engine (2.2) by the fact that

$$(2.10) \qquad [1]\, t_1\, \frac{\partial}{\partial t_1}\, f(t_1, \dots, t_n) = 0,$$

provided that $f(t_1, \dots, t_n)$ has a laurent expansion at $t_1 = 0$. Thus the partial derivative $\partial/\partial t_1 f(t_1, \dots, t_n)$ has no residue at $t_1 = 0$. This is the engine of our machine for the $q = 1$ case.

Recall that we omit q when $q = 1$. Thus

$$(2.11) \qquad \begin{aligned} &bc_{n,m,r}(a, b, k; t_1, \dots, t_n) \\ &= \prod_{i=1}^{n} t_i^{\chi(i \le r)} (1 - t_i)^{a + \chi(n-i+1 \le m)} (1 - \frac{1}{t_i})^a (1 - t_i^2)^b (1 - \frac{1}{t_i^2})^b \\ &\quad \times \prod_{1 \le i < j \le n} (1 - \frac{t_i}{t_j})^k (1 - \frac{t_j}{t_i})^k (1 - t_i t_j)^k (1 - \frac{1}{t_i t_j})^k. \end{aligned}$$

For $m = r = 0$, we have the symmetry

$$(2.12) \qquad bc_n(a, b, k; t_{\pi(1)}^{\epsilon(1)}, \dots, t_{\pi(n)}^{\epsilon(n)}) = bc_n(a, b, k; t_1, \dots, t_n),$$

where

$$(2.13) \qquad \epsilon(i) = \pm 1,\ 1 \le i \le n,\ \text{and}\ \pi \in S_n.$$

We have

$$(2.14) \qquad \begin{aligned} BC_{n,m}(a, b, k) &= [1] \prod_{i=1}^{r} t_i\, bc_n(a, b, k; t_1, \dots, t_n) \\ &= [1] \prod_{i=1}^{r} t_{\pi(i)}^{\epsilon(i)}\, bc_n(a, b, k; t_1, \dots, t_n). \end{aligned}$$

The root systems B_n and C_n have the same Weyl group, namely the semidirect product $2^n \rtimes S_n$ of 2^n by S_n, where 2^n is the group of sign changes in the first n coordinates. We let the signed permutation $\langle \epsilon, \pi \rangle \in 2^n \rtimes S_n$ correspond to the substitution $t_i \to t_{\pi(i)}^{\epsilon(i)}$, $1 \le i \le n$. We call (2.14) the transportation property for BC_n. Since S_n is embedded in $2^n \rtimes S_n$ by $\epsilon(i) = 1$, $1 \le i \le n$, we see that (2.14) explicitly expresses a constant term formulation of the transportation property for A_{n-1} in terms of $bc_n(a,b,k;t_1,\ldots,t_n)$.

The $q = 1$ case of Theorem 4 (1.18) is

$$(2.15) \quad BC_{n,m}(a,b,k) = \prod_{i=1}^{n} \frac{(2b + 2(n-i)k)!}{(b + (n-i)k)!}$$

$$\times \frac{(2a + 2b + 2(n-i)k + \chi(i \le m))!}{(a + b + (n-i)k)!\,(a + 2b + (2n - i - 1)k + \chi(i \le m))!} \frac{(ik)!}{k!}.$$

This may be proved by applying the engine (2.10) to $f(t_1,\ldots,t_n) = bc_{n,m}(a,b,k;t_1,\ldots,t_n)$, $m \ge 1$. Using the symmetry (2.12) of $bc_n(a,b,k;t_1,\ldots,t_n)$ and the transportation property (2.14), we obtain

$$(2.16) \quad \begin{aligned} 0 &= (a + 2b + 1 + (2n - m - 1)k)\,BC_{n,m}(a,b,k) \\ &\quad - (2a + 2b + 1 + 2(n - m)k)\,BC_{n,m-1}(a,b,k), \quad m \ge 1, \end{aligned}$$

or

$$(2.17) \quad BC_{n,m}(a,b,k) = \frac{(2a + 2b + 1 + 2(n - m)k)}{(a + 2b + 1 + (2n - m - 1)k)}\,BC_{n,m-1}(a,b,k), \quad m \ge 1.$$

This establishes the dependence of $BC_{n,m}(a,b,k)$ on m. Repeated applications of (2.17) gives

$$(2.18) \quad BC_{n,m}(a,b,k) = \prod_{i=1}^{m} \frac{(2a + 2b + 1 + 2(n - i)k)}{(a + 2b + 1 + (2n - i - 1)k)}\,BC_n(a,b,k), \quad m \ge 1.$$

Since

$$(2.19) \qquad\qquad (1 - t_i)(1 - \frac{1}{t_i}) = (1 - t_i) + (1 - \frac{1}{t_i}),$$

the transportation property (2.14) gives

$$BC_n(a + 1, b, k) = [1] \prod_{i=1}^{n} (1 - t_i)(1 - \frac{1}{t_i})\,bc_n(a,b,k;t_1,\ldots,t_n)$$

$$(2.20) \qquad\qquad = 2^n\,[1] \prod_{i=1}^{n} (1 - t_i)\,bc_n(a,b,k;t_1,\ldots,t_n)$$

$$= 2^n\,BC_{n,n}(a,b,k).$$

Observe that (2.17) and (2.20) establish the dependence of $BC_{n,m}(a,b,k)$ on m and a. Substituting the $m = n$ case of (2.18) into (2.20) gives

$$(2.21) \qquad BC_n(a + 1, b, k) = 2^n \prod_{i=1}^{n} \frac{(2a + 2b + 1 + 2(n - i)k)}{(a + 2b + 1 + (2n - i - 1)k)}\,BC_n(a,b,k).$$

Thus the parameter m refines the dependence of $BC_n(a, b, k)$ on the parameter a.

Expanding the polynomial $bc_n(0, b, k; t_1, \ldots, t_n)$ and extracting the constant term, we may express $BC_{n,n}(a, b, k)$ as a finite sum, each term of which may be analytically continued as a function of a. Using (1.5), we may evaluate the analytic continuation at $a = -1 - 2b - (n-1)k$. We obtain

(2.22)

$$BC_{n,n}(-1 - 2b - (n-1)k, b, k) = (-1)^{(nb+\binom{n}{2}k)} [1] \prod_{1 \le i < j \le n} (1 - \frac{t_i}{t_j})^k (1 - \frac{t_j}{t_i})^k$$

$$= (-1)^{(nb+\binom{n}{2}k)} \frac{(nk)!}{(k!)^n}.$$

The $q = 1$ case (2.15) of Theorem 4 follows by follows by analytically continuing the product on the right side of (2.15) as a function of a and comparing the result for $m = n$, $a = -1 - 2b - (n-1)k$.

Expanding $bc_{n,m,r}(a, b, k; t_1, \ldots, t_n)$ by $t_r = 1 - (1 - t_r)$ and using the transportation property (2.14), we have

(2.23) $$BC_{n,m,r}(a, b, k) = BC_{n,m,r-1}(a, b, k) - BC_{n,m+1,r-1}(a, b, k), \quad r \ge 1.$$

One may now show by induction on r that

(2.24)
$$0 = (a + 2b + 1 + (2n - m - r - 1)k) \, BC_{n,m,r}(a, b, k)$$
$$- (2a + 2b + 1 + 2(n - m)k) \, BC_{n,m-1,r}(a, b, k)$$
$$- rk \, BC_{n,m,r-1}(a, b, k), \quad m \ge 1.$$

For $r = 0$, this is (2.16). To do the induction, take (2.24) minus (2.24) with m replaced by $m + 1$ and simplify using (2.23). We may rewrite (2.23) as

(2.25) $$BC_{n,m+1,r-1}(a, b, k) = BC_{n,m,r-1}(a, b, k) - BC_{n,m,r}(a, b, k), \quad r \ge 1.$$

Replace m, r, by $m + 1$, $r - 1$, in (2.24) and apply (2.25) to the first and last term. Changing sign, we obtain

(2.26)
$$0 = (a + 2b + 1 + (2n - m - r - 1)k) \, BC_{n,m,r}(a, b, k)$$
$$+ (a - mk) \, BC_{n,m,r-1}(a, b, k)$$
$$+ (r - 1)k \, BC_{n,m,r-2}(a, b, k), \quad r \ge 1.$$

For $m = 0$, this is

(2.27)
$$0 = (a + 2b + 1 + (2n - r - 1)k) \, BC_{n,0,r}(a, b, k)$$
$$+ a \, BC_{n,0,r-1}(a, b, k)$$
$$+ (r - 1)k \, BC_{n,0,r-2}(a, b, k), \quad r \ge 1.$$

Alternatively, (2.27) follows by applying the engine (2.10) to $f(t_1, \ldots, t_n) = bc_{n,0,r}(a, b, k; t_1, \ldots, t_n)$, $r \ge 1$. We may obtain (2.26) by induction on m and

(2.24) and (2.17) then follow by simple rearrangements. Since this approach is taken in the proof of Theorem 4, we outline the basic steps. Simple calculations give

(2.28)

$$s \frac{\partial}{\partial s} \left((1 - \frac{s}{t})^k (1 - \frac{t}{s})^k \right) = k \frac{(s+t)}{(s-t)} (1 - \frac{s}{t})^k (1 - \frac{t}{s})^k,$$

(2.29)

$$s \frac{\partial}{\partial s} \left((1 - s)^a (1 - \frac{1}{s})^a \right) = a \frac{(s+1)}{(s-1)} (1 - s)^a (1 - \frac{1}{s})^a,$$

(2.30)

$$s \frac{\partial}{\partial s} \left((1 - st)^k (1 - \frac{1}{st})^k \right) = k \frac{(st+1)}{(st-1)} (1 - st)^k (1 - \frac{1}{st})^k,$$

(2.31)

$$s \frac{\partial}{\partial s} \left((1 - s^2)^b (1 - \frac{1}{s^2})^b \right) = 2b \frac{(s^2+1)}{(s^2-1)} (1 - s^2)^b (1 - \frac{1}{s^2})^b.$$

Taking $s = t_1$ and $t = t_v$, $2 \le v \le n$, in (2.28)–(2.31), we have

(2.32)

$$t_1 \frac{\partial}{\partial t_1} \left(bc_n(a, b, k; t_1, \ldots, t_n) \right)$$

$$= \left(k \sum_{v=2}^{n} \frac{(t_1 + t_v)}{(t_1 - t_v)} + a \frac{(t_1 + 1)}{(t_1 - 1)} + k \sum_{v=2}^{n} \frac{(t_1 t_v + 1)}{(t_1 t_v - 1)} + 2b \frac{(t_1^2 + 1)}{(t_1^2 - 1)} \right)$$

$$\times \, bc_n(a, b, k; t_1, \ldots, t_n).$$

Observe that the factors $(t_1 + t_v)/(t_1 - t_v)$, $(t_1 + 1)/(t_1 - 1)$, $(t_1 t_v + 1)/(t_1 t_v - 1)$ and $(t_1^2 + 1)/(t_1^2 - 1)$, which arise in the partial derivative (2.32), are antisymmetric under $t_1 \leftrightarrow t_v$, $t_1 \leftrightarrow 1/t_1$, $t_1 \leftrightarrow 1/t_v$ and $t_1 \leftrightarrow 1/t_1$, respectively. We view these substitutions as corresponding to the reflections σ_α along α where we differentiate $(1 - e^\alpha)^{k\alpha} (1 - e^{-\alpha})^{k\alpha}$.

By the engine (2.10), the constant term in (2.32) is 0. Using the symmetry (2.12) of $bc_n(a, b, k; t_1, \ldots, t_n)$, we see that each term occurring in (2.32) is antisymmetric under the corresponding substitution and hence has constant term 0.

Since t_1 is a factor of $\prod_{i=1}^{r} t_i$ when $r \geq 1$, we have

$$
\begin{aligned}
t_1 \frac{\partial}{\partial t_1} &\left(\prod_{i=1}^{r} t_i \, bc_n(a,b,k;t_1,\dots,t_n) \right) \\
&= k \sum_{v=2}^{n} \frac{(t_1 + t_v)}{(t_1 - t_v)} \prod_{i=1}^{r} t_i \, bc_n(a,b,k;t_1,\dots,t_n) \\
&\quad + a \frac{(t_1 + 1)}{(t_1 - 1)} \prod_{i=1}^{r} t_i \, bc_n(a,b,k;t_1,\dots,t_n) \\
&\quad + k \sum_{v=2}^{n} \frac{(t_1 t_v + 1)}{(t_1 t_v - 1)} \prod_{i=1}^{r} t_i \, bc_n(a,b,k;t_1,\dots,t_n) \\
&\quad + 2b \frac{(t_1^2 + 1)}{(t_1^2 - 1)} \prod_{i=1}^{r} t_i \, bc_n(a,b,k;t_1,\dots,t_n) \\
&\quad + \prod_{i=1}^{r} t_i \, bc_n(a,b,k;t_1,\dots,t_n), \quad r \geq 1.
\end{aligned}
$$

(2.33)

Applying the engine (2.10) to $f(t_1,\dots,t_n) = bc_{n,0,r}(a,b,k;t_1,\dots,t_n)$, $r \geq 1$, we have

$$
\begin{aligned}
0 = &\ k \sum_{v=2}^{n} A_{n,0,r}^{v}(a,b,k) + a \, E_{n,0,r}(a,b,k) \\
&+ k \sum_{v=2}^{n} K_{n,0,r}^{v}(a,b,k) + 2b \, F_{n,0,r}(a,b,k) \\
&+ Z_{n,0,r}(a,b,k), \quad r \geq 1,
\end{aligned}
$$

(2.34)

where

(2.35)

$$A_{n,0,r}^v(a,b,k) = [1]\frac{(t_1+t_v)}{(t_1-t_v)}\prod_{i=1}^r t_i\, bc_n(a,b,k;t_1,\dots,t_n),\quad 2\le v\le n,$$

(2.36)

$$E_{n,0,r}(a,b,k) = [1]\frac{(t_1+1)}{(t_1-1)}\prod_{i=1}^r t_i\, bc_n(a,b,k;t_1,\dots,t_n),$$

(2.37)

$$K_{n,0,r}^v(a,b,k) = [1]\frac{(t_1t_v+1)}{(t_1t_v-1)}\prod_{i=1}^r t_i\, bc_n(a,b,k;t_1,\dots,t_n),\quad 2\le v\le n,$$

(2.38)

$$F_{n,0,r}(a,b,k) = [1]\frac{(t_1^2+1)}{(t_1^2-1)}\prod_{i=1}^r t_i\, bc_n(a,b,k;t_1,\dots,t_n),$$

(2.39)

$$Z_{n,0,r}(a,b,k) = [1]\prod_{i=1}^r t_i\, bc_n(a,b,k;t_1,\dots,t_n).$$

We may use the antisymmetries of the terms occurring in the partial derivative (2.32) to remove the factors occurring in the denominators of the functions on the right sides of (2.35)–(2.38) as follows. The constant term in each function is fixed by the corresponding substitution. Averaging the two values, we obtain

(2.40) $$A_{n,0,r}^v(a,b,k) = 0,\quad 2\le v\le r,$$

(2.41)
$$A_{n,0,r}^v(a,b,k) = \frac{1}{2}[1]\frac{(t_1+t_v)}{(t_1-t_v)}(t_1-t_v)\prod_{i=2}^r t_i\, bc_n(a,b,k;t_1,\dots,t_n)$$
$$= \frac{1}{2}[1](t_1+t_v)\prod_{i=2}^r t_i\, bc_n(a,b,k;t_1,\dots,t_n),\quad r< v\le n,$$

(2.42)
$$E_{n,0,r}(a,b,k) = \frac{1}{2}[1]\frac{(t_1+1)}{(t_1-1)}(t_1-\frac{1}{t_1})\prod_{i=2}^r t_i\, bc_n(a,b,k;t_1,\dots,t_n)$$
$$= \frac{1}{2}[1]\frac{(t_1+1)^2}{t_1}\prod_{i=2}^r t_i\, bc_n(a,b,k;t_1,\dots,t_n),$$

(2.43)

$$K_{n,0,r}^v(a,b,k) = \frac{1}{2}\,[1]\,\frac{(t_1 t_v + 1)}{(t_1 t_v - 1)}\,(t_1 t_v - \frac{1}{t_1 t_v})\,\prod_{\substack{i=2 \\ i \neq v}}^{r} t_i\; bc_n(a,b,k;t_1,\dots,t_n)$$

$$= \frac{1}{2}\,[1]\,\frac{(t_1 t_v + 1)^2}{t_1 t_v}\,\prod_{\substack{i=2 \\ i \neq v}}^{r} t_i\; bc_n(a,b,k;t_1,\dots,t_n),\quad 2 \leq v \leq r,$$

(2.44)

$$K_{n,0,r}^v(a,b,k) = \frac{1}{2}\,[1]\,\frac{(t_1 t_v + 1)}{(t_1 t_v - 1)}\,(t_1 - \frac{1}{t_v})\,\prod_{i=2}^{r} t_i\; bc_n(a,b,k;t_1,\dots,t_n)$$

$$= \frac{1}{2}\,[1]\,\frac{(t_1 t_v + 1)}{t_v}\,\prod_{i=2}^{r} t_i\; bc_n(a,b,k;t_1,\dots,t_n),\quad r < v \leq n,$$

(2.45)

$$F_{n,0,r}(a,b,k) = \frac{1}{2}\,[1]\,\frac{(t_1^2 + 1)}{(t_1^2 - 1)}\,(t_1 - \frac{1}{t_1})\,\prod_{i=2}^{r} t_i\; bc_n(a,b,k;t_1,\dots,t_n)$$

$$= \frac{1}{2}\,[1]\,\frac{(t_1^2 + 1)}{t_1}\,\prod_{i=2}^{r} t_i\; bc_n(a,b,k;t_1,\dots,t_n).$$

Using the transportation property (2.14), we have

(2.46)

$$A_{n,0,r}^v(a,b,k) = BC_{n,0,r}(a,b,k),\quad r < v \leq n.$$

(2.47)

$$E_{n,0,r}(a,b,k) = BC_{n,0,r}(a,b,k) + BC_{n,0,r-1}(a,b,k),$$

(2.48)

$$K_{n,0,r}^v(a,b,k) = BC_{n,0,r-2}(a,b,k) + BC_{n,0,r}(a,b,k),\quad 2 \leq v \leq r,$$

(2.49)

$$K_{n,0,r}^v(a,b,k) = BC_{n,0,r}(a,b,k),\quad r < v \leq n,$$

(2.50)

$$F_{n,0,r}(a,b,k) = BC_{n,0,r}(a,b,k).$$

Observe that

(2.51)

$$Z_{n,0,r}(a,b,k) = BC_{n,0,r}(a,b,k).$$

Substituting (2.40) and (2.46)–(2.51) into (2.34) and simplifying yields (2.27). Using (2.25), we may establish (2.26) by induction on m. Simple rearrangements give (2.24) which includes (2.17) as the $r = 0$ case.

Since the positive and negative roots are treated differently in $_q bc_n(a,b,k;t_1,\dots,t_n)$, the symmetry (2.12) does not extend to the q-case. As in the proof [Ka3] of

Askey's q-Selberg integral, where the symmetry (2.4) does not extend, this is the major impediment to the proof of Theorem 4. However, $_qbc_n(a,b,k;t_1,\ldots,t_n)$ satisfies certain near symmetries which are just what we require to do the analysis of the proof of Theorem 4 locally. The centerpiece of our proof is a q-analogue of the transportation property (2.14) which we call the q-transportation theory for BC_n. This is proved locally by using the basic property of the simple reflections of B_n and C_n that the reflection along a simple positive root $\alpha \in R^+$ sends α to $-\alpha$ and permutes the other positive roots.

In Section 3, we review the role that simple roots and reflections play in the relationships between the root systems A_{n-1}, D_n, B_n, C_n and BC_n. We explicitly express the basic property of the simple reflections and the geometry of the simple roots of B_n and C_n in terms of $_qbc_n(a,b,k;t_1,\ldots,t_n)$. Lemma 5 gives certain near symmetries of $_qbc_n(a,b,k;t_1,\ldots,t_n)$ which explicitly express the basic property of the simple reflections of B_n and C_n. Lemma 6 gives the effects of certain substitutions on $_qbc_n(a,b,k;t_1,\ldots,t_n)$ which explicitly express the geometry of the simple roots of B_n and C_n.

In Section 4, we introduce the partial q-derivative and show that if $f(t_1,\ldots,t_n)$ has a Laurent expansion at $t_1 = 0$, then $\partial_q/\partial_q t_1 f(t_1,\ldots,t_n)$ has no residue at $t_1 = 0$. This q-analogue of (2.10) is the q-engine of our q-machine. It is equivalent to the fact that if $f(t_1,\ldots,t_n)$ has a Laurent expansion at $t_1 = 0$, then the constant term $[1]f(t_1,\ldots,t_n)$ is fixed by $t_1 \to qt_1$. Lemma 7 gives an identity for the partial q-derivative of $_qbc_n(a,b,k;t_1,\ldots,t_n)$ and shows that each term occurring therein is antisymmetric under a certain substitution. This is a q-analogue of (2.32). We apply the q-engine to $f(t_1,\ldots,t_n) = {}_qbc_{n,0,r}(a,b,k;t_1,\ldots,t_n)$, $r \geq 1$, and obtain the q-analogue Lemma 8 of (2.34).

In Section 5, we use the antisymmetries of Lemma 7 to remove the factors occurring in the denominators of the terms of Lemma 8. We obtain Lemma 9 which gives q-analogues of (2.40)–(2.45).

In Section 6, we use the basic property of the simple reflections of B_n and C_n to give a local proof of the q-transportation theory for BC_n. The global version of the integral formulation of the q-transportation theory for A_{n-1} is the q-analogue [Ka2, (4.14)] of (2.5). This was used in the proof [Ka3] of Askey's q-Selberg integral. Lemma 10 is the local version of the constant term formulation. Lemma 11, which we call the q-transportation theory for A_{n-1}, explicitly expresses Lemma 10 in terms of $_qbc_n(a,b,k;t_1,\ldots,t_n)$. Lemmas 12 and 14 extend the q-transportation theory for A_{n-1} to B_n and C_n, respectively. Lemmas 13 and 15 explicitly express the results in terms of $_qbc_n(a,b,k;t_1,\ldots,t_n)$. Lemmas 11, 13 and 15, which give q-analogues of the transportation property (2.14), constitute the q-transportation theory for BC_n.

In Section 7, we complete the evaluation of the constant terms occurring in Lemma 8 by applying the q-transportation theory for C_n, Lemmas 11 and 15, to the results of Lemma 9. We require Corollary 16, which is a technical corollary of Lemmas 11 and 15. We obtain Lemma 17 which gives q-analogues of (2.40) and (2.46)–(2.51).

In Section 8, we give q-analogues of some functional equations which establish the dependence of $_qBC_{n,m}(a,b,k)$ on m and a. Substituting the results of Lemma 17 into Lemma 8 gives the q-analogue Lemma 18 of (2.27). Using the q-transportation theory for A_{n-1}, Lemma 11, we rearrange Lemma 18, obtaining Lemmas 19, 20 and 21, which give q-analogues of (2.26), (2.24) and (2.17), respectively. Using the q-transportation theory for B_n, Lemmas 11 and 13, we obtain the q-analogue Lemma 22 of (2.20). Lemmas 21 and 22 establish the dependence of $_qBC_{n,m}(a,b,k)$ on m and a.

In Section 9, we show how to obtain the q-analogues Lemmas 18 and 21 of (2.27) and (2.17), respectively, using only the q-transportation theory for BC_n, Lemmas 11, 13 and 15. This is based upon the fact that B_n and C_n have the same Weyl group. Since Lemmas 21 and 22 are the only results which we require to complete the proof of Theorem 4 (1.18), we may supplant the q-engine of our q-machine by the q-transportation theory for BC_n.

In Section 10, we complete the proof of Theorem 4 (1.18). We analytically continue $_qBC_{n,n}(a,b,k)$ as a function of a and, using the $a=b=0$ case of Theorem 3 (1.14), we obtain a q-analogue of (2.22). Theorem 4 follows by analytically continuing the product on the right side of (1.18) as a function of a and comparing the result for $m=n$, $a=-1-2b-(n-1)k$.

In Section 11, we prove Lemma 23 by which we may evaluate $_qBC_{n,m,r}(a,b,k)$.

In Section 12, we discuss the q-Macdonald-Morris conjecture for B_n, $B_n^\vee$, C_n, $C_n^\vee$ and D_n. We observe that the B_n, $B_n^\vee$ and D_n cases of the conjecture follow from Theorem 4 (1.18). We give some of the details for C_n and $C_n^\vee$ and acknowledge the priority of Gustafson's recent proof [Gu1] of these cases of the conjecture. This illustrates the basic steps required to apply our methods to the conjecture when the reduced irreducible root system R does not have a minuscule weight.

We give [Ka4] a simple proof of an Aomoto type extension of the q-Morris theorem. Gustafson [Gu1] incorporates the Askey-Wilson integral [AW1] into an elegant multivariable q-Selberg integral which gives the q-Macdonald-Morris conjecture for B_n, $B_n^\vee$, C_n, $C_n^\vee$, D_n and BC_n. We give [Ka5] a simple proof of an Aomoto type extension of Gustafson's theorem.

Let R be a reduced, irreducible root system of an isolated type. We may give an identity for a partial q-derivative which allows us to use the q-engine of our q-machine. If R does not have a minuscule weight, then it has a quasi-minuscule weight and we may follow the approach given in Section 12 for C_n and $C_n^\vee$. The recent work of Garvan [Ga1, Ga2] and Garvan and Gonnet [GG1] suggests that there exist Aomoto type extensions of the q-Macdonald-Morris conjecture for R. We should be able to use our q-machine to prove the required functional equations. Following [Ka4, Ka5], we should be able to give a simple proof, which is read off from the identity for a partial q-derivative. The author has recently carried out these calculations by hand for G_2.

Hopefully there is a general formula for an Aomoto type extension of the q-Macdonald-Morris conjecture for R which can be expressed in terms of the geometry or algebra of R and a classification free proof is not far behind.

3. The simple roots and reflections of B_n and C_n

In this section, we review the role that simple roots and reflections play in the relationships between the root systems A_{n-1}, D_n, B_n, C_n and BC_n. We explicitly express the basic property of the simple reflections and the geometry of the simple roots of B_n and C_n in terms of $_q bc_n(a, b, k; t_1, \ldots, t_n)$. Lemma 5 gives certain near symmetries of $_q bc_n(a, b, k; t_1, \ldots, t_n)$ which explicitly express the basic property of the simple reflections of B_n and C_n. Lemma 6 gives the effects of certain substitutions on $_q bc_n(a, b, k; t_1, \ldots, t_n)$ which explicitly express the geometry of the simple roots of B_n and C_n.

Let V be a vector space over the reals with dimension ℓ, positive definite symmetric bilinear form (α, β), and orthonormal basis $\{e_1, \ldots, e_\ell\}$. Let R be a root system on V. We use many of the properties of root systems given in Grove and Benson [GB1], Carter [Ca1] and Humphreys [Hu1]. R is finite, spans V and does not contain 0. The reflection of $v \in V$ along $\alpha \in R$ is given by

$$(3.1) \qquad \sigma_\alpha(v) = v - 2 \frac{(\beta, \alpha)}{(\alpha, \alpha)} \alpha, \quad v \in V, \ \alpha \in R.$$

The reflection σ_α is an involution which sends α to $-\alpha$, fixes the hyperplane perpendicular to α

$$(3.2) \qquad P_\alpha = \{v \in V \mid (v, \alpha) = 0\}$$

pointwise, and preserves length and inner product

$$(3.3) \qquad |\sigma_\alpha(v)| = |v|, \ (\sigma_\alpha(u), \sigma_\alpha(v)) = (u, v), \quad u, v \in V, \ \alpha \in R.$$

Thus σ_α is an orthogonal transformation.

Each reflection σ_α, $\alpha \in R$, acts transitively on the roots of a given length. The reflections σ_α, $\alpha \in R$, generate a finite group W called the Weyl group of R. Since each $w \in W$ is a product of reflections and each of these is an orthogonal transformation, we have that w is an orthogonal transformation. Thus w preserves length and inner product

$$(3.4) \qquad |w(v)| = |v|, \ (w(u), w(v)) = (u, v), \quad u, v \in V, \ w \in W.$$

Since it plays such an important role, we set

$$(3.5) \qquad \langle \beta, \alpha \rangle = 2 \frac{(\beta, \alpha)}{(\alpha, \alpha)}.$$

Observe that $\langle \beta, \alpha \rangle$ is linear only in the first variable. The angles between the roots of R are severely restricted by the requirement that

$$(3.6) \qquad \langle \beta, \alpha \rangle \text{ is an integer whenever } \alpha, \beta \in R.$$

Let R^+ be a system of positive roots. Half of the sum of the positive roots is given by

$$(3.7) \qquad \rho = \frac{1}{2} \sum_{\alpha \in R^+} \alpha$$

16

and we may recover R^+ from ρ by the geometry

$$(3.8) \qquad R^+ = \{\alpha \in R \mid (\alpha, \rho) > 0\}$$

of the positive roots.

Let R be a reduced root system. This requires that

$$(3.9) \qquad \text{if } \alpha, \, c\alpha \in R, \text{ then } c = \pm 1.$$

The base $\mathcal{B}(R^+) = \{\alpha_1, \dots, \alpha_\ell\}$ of simple positive roots corresponding to R^+ is chosen so that the inner products (α_i, ρ), $1 \le i \le \ell$, are as small as possible. Thus the geometry of the simple roots is given by

$$(3.10) \qquad (\alpha, \rho) > (\alpha_i, \rho) \text{ whenever } 1 \le i \le \ell \text{ and } \alpha \in R^+ - \{\alpha_1, \dots, \alpha_\ell\}.$$

We denote the simple reflections by $\sigma_i = \sigma_{\alpha_i}$, $1 \le i \le \ell$. The basic property of the simple reflections is that each σ_i, $1 \le i \le \ell$, changes the sign of α_i and permutes the other positive roots. Thus

$$(3.11) \qquad \sigma_i(\alpha_i) = -\alpha_i, \; \sigma_i(R^+ - \{\alpha_i\}) = R^+ - \{\alpha_i\}, \quad 1 \le i \le \ell.$$

The Weyl group W of R is generated by the simple reflections

$$(3.12) \qquad W = \langle \sigma_1, \dots, \sigma_\ell \rangle.$$

By the geometry (3.8) of the positive roots and the orthogonality (3.4) of w, we see that the image of the system R^+ of positive roots is the system of positive roots which corresponds to the image of ρ. Thus we have

$$(3.13) \qquad w(R^+) = \{\alpha \in R \mid (\alpha, w(\rho)) > 0\}, \quad w \in W.$$

The orthogonality (3.4) of w gives

$$(3.14) \qquad (\alpha_i, \rho) = (w(\alpha_i), w(\rho)), \quad w \in W.$$

By the geometry (3.10) of the simple roots and the orthogonality (3.4) of w, we see that the image of the base $\mathcal{B}(R^+)$ of simple positive roots corresponding to R^+ is the base which corresponds to the image of R^+. Thus we have

$$(3.15) \qquad \begin{aligned} \mathcal{B}(w(R^+)) &= \{w(\alpha_1), \dots, w(\alpha_\ell)\} \\ &= w(\mathcal{B}(R^+)). \end{aligned}$$

Let the reduced root system S be an extension of R with rank $\ell + 1$. Choosing S^+ to contain R^+, we have

$$(3.16) \qquad \mathcal{B}(S^+) = \mathcal{B}(R^+) \bigcup \{\alpha_{\ell+1}\}$$

and we let $\sigma_{\ell+1} = \sigma_{\alpha_{\ell+1}}$. Since each α_i, $1 \leq i \leq \ell$, is a simple positive root of both R and S, we see that each σ_i, $1 \leq i \leq \ell$, permutes $R^+ - \{\alpha_i\}$ and $S^+ - \{\alpha_i\}$. Hence it permutes the difference $S^+ - R^+$. Thus we have

$$(3.17) \qquad \sigma_i(S^+ - R^+) = S^+ - R^+, \quad 1 \leq i \leq \ell.$$

If $\alpha_{\ell+1}$ has a new root length, then $\sigma_{\ell+1}$ permutes R^+. Thus we have that

$$(3.18) \qquad \text{if } |\alpha_{\ell+1}| \neq |\alpha| \text{ for all } \alpha \in R, \text{ then } \sigma_{\ell+1}(R^+) = R^+.$$

Taking A_{n-1}^+ as in (1.4), we associate the function

$$(3.19) \qquad {}_qa_{n-1}(k; t_1, \dots, t_n) = \prod_{1 \leq i < j \leq n} (\frac{t_i}{t_j})_k (q\frac{t_j}{t_i})_k$$

with A_{n-1}. The simple positive roots of A_{n-1} are given by

$$(3.20) \qquad \alpha_{v-1} = e_{v-1} - e_v, \quad 2 \leq v \leq n.$$

Since $\sigma_{v-1}(e_v) = e_{v-1}$ and $\sigma_{v-1}(e_{v-1}) = e_v$, the simple reflections are the adjacent transpositions $e_{v-1} \leftrightarrow e_v$, $2 \leq v \leq n$, and the corresponding substitutions are given by

$$(3.21) \qquad t_{v-1} \leftrightarrow t_v, \quad 2 \leq v \leq n.$$

We view the Weyl group S_n of A_{n-1} as the substitutions $t_i \to t_{\pi(i)}$, $1 \leq i \leq n$, where $\pi \in S_n$.

Define the function ${}_qa_{n-1}^v(k; t_1, \dots, t_n)$ by

$$(3.22) \quad {}_qa_{n-1}(k; t_1, \dots, t_n) = (\frac{t_{v-1}}{t_v})_k (q\frac{t_v}{t_{v-1}})_k \; {}_qa_{n-1}^v(k; t_1, \dots, t_n), \quad 2 \leq v \leq n.$$

Observe that ${}_qa_{n-1}^v(k; t_1, \dots, t_n)$ equals

$$(3.23) \quad \prod_{1 \leq i < v-1} (\frac{t_i}{t_{v-1}})_k (q\frac{t_{v-1}}{t_i})_k (\frac{t_i}{t_v})_k (q\frac{t_v}{t_i})_k \prod_{v < j \leq n} (\frac{t_{v-1}}{t_j})_k (q\frac{t_j}{t_{v-1}})_k (\frac{t_v}{t_j})_k (q\frac{t_j}{t_v})_k$$

times a function which is independent of t_{v-1} and t_v. Thus ${}_qa_{n-1}^v(k; t_1, \dots, t_n)$ is symmetric
$$(3.24)$$
$${}_qa_{n-1}^v(k; t_1, \dots, t_n) = {}_qa_{n-1}^v(k; t_1, \dots, t_{v-2}, t_v, t_{v-1}, t_{v+1}, \dots, t_n), \quad 2 \leq v \leq n,$$

in t_{v-1} and t_v. The symmetry (3.24) explicitly expresses the basic property of the simple reflections of A_{n-1}.

Apply the simple reflections σ_{v-1} to A_{n-1}^+ with v running from 2 to n. Set

$$(3.25) \qquad w_v = \sigma_{v-1} \cdots \sigma_1, \quad 1 \leq v \leq n.$$

We use the convention that $a_i, \ldots, a_{i-1}$ is an empty sequence. Taking the empty product to be unity, we have $w_1 = 1_n$, where 1_n is the identity permutation in S_n. We have

$$(3.26) \qquad w_{v-1}(e_1, \ldots, e_n) = (e_2, \ldots, e_{v-1}, e_1, e_v, \ldots, e_n), \quad 2 \le v \le n+1.$$

By the geometry (3.15) of the simple roots of A_{n-1}, we see that

$$(3.27) \qquad e_1 - e_v = w_v(e_{v-1} - e_v) \in \mathcal{B}(w_v(A_{n-1}^+)), \quad 2 \le v \le n.$$

Observe that
$$(3.28)$$
$$_q a_{n-1}(k; t_2, \ldots, t_{v-1}, t_1, t_v, \ldots, t_n)$$
$$= \prod_{i=2}^{v-1} (q\frac{t_1}{t_i})_k (\frac{t_i}{t_1})_k \prod_{j=v}^{n} (\frac{t_1}{t_j})_k (q\frac{t_j}{t_1})_k \prod_{1 \le i < j \le n} (\frac{t_i}{t_j})_k (q\frac{t_j}{t_i})_k, \quad 2 \le v \le n+1.$$

This explicitly expresses the geometry of the simple roots of A_{n-1} in terms of $_q a_{n-1}(k; t_1, \ldots, t_n)$.

We may extend A_{n-1} to D_n by adding the simple positive root

$$(3.29) \qquad \alpha_n = e_{n-1} + e_n.$$

This gives the system D_n^+ (1.4) of positive roots, to which we associate the function

$$(3.30) \qquad _q d_n(k; t_1, \ldots, t_n) = \prod_{1 \le i < j \le n} (\frac{t_i}{t_j})_k (q\frac{t_j}{t_i})_k (t_i t_j)_k (\frac{q}{t_i t_j})_k.$$

We have

$$(3.31) \qquad _q d_n(k; t_1, \ldots, t_n) = {}_q ex_n(k; t_1, \ldots, t_n) {}_q a_{n-1}(k; t_1, \ldots, t_n),$$

where

$$(3.32) \qquad _q ex_n(k; t_1, \ldots, t_n) = \prod_{1 \le i < j \le n} (t_i t_j)_k (\frac{q}{t_i t_j})_k.$$

Observe that in accordance with (3.17) the function $_q ex_n(k; t_1, \ldots, t_n)$ associated with the roots of D_n which are not in A_{n-1}, that is with $D_n - A_{n-1}$, is symmetric

$$(3.33) \qquad _q ex_n(k; \pi(t_1), \ldots, \pi(t_n)) = {}_q ex_n(k; t_1, \ldots, t_n), \quad \pi \in S_n,$$

in $t_1, \ldots, t_n$.

The substitution corresponding to the simple reflection $\sigma_n = \sigma_{\alpha_n}$ along α_n given by (3.29) is $t_{n-1} \leftrightarrow 1/t_n$. The Weyl group of D_n is the semidirect product $2^{n-1} \rtimes S_n$ of 2^{n-1} by S_n, where 2^{n-1} is the group of sign changes in an even number of the first n coordinates. We let the signed permutation $\langle \epsilon, \pi \rangle \in 2^{n-1} \rtimes S_n$ correspond to the substitution $t_i \to t_{\pi(i)}^{\epsilon(i)}$, $1 \le i \le n$.

Observe that

$$f(s,t) = (\frac{s}{t})_k(q\frac{t}{s})_k(st)_k(\frac{q}{st})_k$$

(3.34)
$$= f(\frac{q}{s},t)$$

$$= f(s,\frac{1}{t}).$$

If $s = t_i$, $t = t_j$, where $1 \le i < j \le n$, then we can only have $s = t_1$ and $t = t_n$. Hence we have the symmetries

(3.35)
$$_qd_n(k;t_1,\dots,t_n) = {}_qd_n(k;t_1,\dots,t_{n-1},\frac{1}{t_n})$$

$$= {}_qd_n(k;\frac{q}{t_1},t_2\dots,t_n).$$

We may extend A_{n-1} to B_n or C_n by adding the simple positive root

(3.36)
$$\alpha_n = e_n \text{ or } \alpha_n = 2e_1,$$

respectively. The corresponding simple reflections are the sign changes $e_n \leftrightarrow -e_n$ or $e_1 \leftrightarrow -e_1$, respectively. The root systems B_n and C_n have the Weyl group, namely the group $2^n \rtimes S_n$ of signed permutations of the first n coordinates.

All of the roots of D_n have length $\sqrt{2}$. The simple, positive root α_n of B_n or C_n given by (3.36) has length 1 or 2, respectively. Hence by (3.18), the simple reflection $\sigma_n = \sigma_{\alpha_n}$ along α_n permutes D_n^+. Thus we have

(3.37)
$$\sigma_n(D_n^+) = D_n^+ \text{ for } B_n \text{ and } C_n.$$

Let the substitutions corresponding to the simple reflections σ_n be given by

(3.38)
$$t_n \to \frac{1}{t_n} \text{ and } t_1 \to \frac{q}{t_1},$$

respectively. Then (3.37) is explicitly expressed by the symmetries (3.35) of $_qd_n(k; t_1,\dots,t_n)$.

Since

(3.39)
$$f(s,t) = f(\frac{\sqrt{q}}{t},\frac{\sqrt{q}}{s})$$

$$= f(-s,-t),$$

we have the symmetries

(3.40)
$$_qd_n(k;t_1,\dots,t_n) = {}_qd_n(k;\frac{\sqrt{q}}{t_n},\dots,\frac{\sqrt{q}}{t_1})$$

$$= {}_qd_n(k;-t_1,\dots,-t_n).$$

Using the substitution $t_i \to \sqrt{q}/t_{n-i+1}$, $1 \le i \le n$, we see that $t_n \leftrightarrow 1/t_n$ becomes $\sqrt{q}/t_1 \leftrightarrow t_1/\sqrt{q}$ or $t_1 \leftrightarrow q/t_1$. Thus the two symmetries (3.35) of $_qd_n(k; t_1, \ldots, t_n)$ are interchangeable. This explicitly expresses the fact that the simple reflection which extends A_{n-1} to B_n or C_n may be taken to correspond to the sign change in the first or last coordinate.

The root system BC_n is formed by combining the reduced root systems B_n and C_n in a manner that respects the geometry of the simple roots of each of the subsystems. We associate the function

$$(3.41) \qquad {}_q\text{short}_n(a; t_1, \ldots, t_n) = \prod_{i=1}^{n} (t_i)_a \left(\frac{q}{t_i}\right)_a$$

with the short roots $B_n - D_n$ of BC_n and the function

$$(3.42) \qquad {}_q\text{long}_n(b; t_1, \ldots, t_n) = \prod_{i=1}^{n} (qt_i^2; q^2)_b \left(\frac{q}{t_i^2}; q^2\right)_b$$

with the long roots $C_n - D_n$ of BC_n. Observe that the parameters a, b and k are associated with roots of length 1, 2 and $\sqrt{2}$, respectively. In accordance with (3.17) we have the symmetries

$$(3.43) \qquad \begin{aligned} {}_q\text{short}_n(a; \pi(t_1), \ldots, \pi(t_n)) &= {}_q\text{short}_n(a; t_1, \ldots, t_n), \quad \pi \in S_n, \\ {}_q\text{long}_n(b; \pi(t_1), \ldots, \pi(t_n)) &= {}_q\text{long}_n(b; t_1, \ldots, t_n), \quad \pi \in S_n, \end{aligned}$$

and, as suggested by (3.18), we have the symmetries

$$(3.44) \qquad \begin{aligned} {}_q\text{short}_n\left(a; \frac{q}{t_1}, t_2, \ldots, t_n\right) &= {}_q\text{short}_n(a; t_1, \ldots, t_n), \\ {}_q\text{long}_n\left(b; t_1, \ldots, t_{n-1}, \frac{1}{t_n}\right) &= {}_q\text{long}_n(b; t_1, \ldots, t_n). \end{aligned}$$

Observe that (3.44) explicitly expresses the compatibility of the geometries of the simple roots of B_n and C_n which allows us to combine them to form the non reduced root system $BC_n = B_n \bigcup C_n$.

These symmetries allow us to incorporate the geometry of the simple roots and the basic property of the simple reflections of B_n and C_n into the elegant creation of Morris

$$(3.45) \qquad \begin{aligned} {}_qbc_n(a, b, k; t_1, \ldots, t_n) &= {}_q\text{short}_n(a; t_1, \ldots, t_n)\, {}_q\text{long}_n(b; t_1, \ldots, t_n) \\ &\times {}_qd_n(k; t_1, \ldots, t_n). \end{aligned}$$

The following lemma explicitly expresses the basic property of the simple reflections of B_n and C_n in terms of $_qbc_n(a, b, k; t_1, \ldots, t_n)$.

Lemma 5. *Define the functions $_q\alpha_n^v(a,b,k;t_1,\ldots,t_n)$, $_q\epsilon_n(a,b,k;t_1,\ldots,t_n)$ and $_qf_n(a,b,k;t_1,\ldots,t_n)$ by*
(3.46)
$$_qbc_n(a,b,k;t_1,\ldots,t_n) = (\frac{t_v}{t_{v-1}})_k(q\frac{t_{v-1}}{t_v})_k \,_q\alpha_n^v(a,b,k;t_1,\ldots,t_n), \quad 2 \le v \le n,$$

$$= (t_n)_a(\frac{q}{t_n})_a \,_q\epsilon_n(a,b,k;t_1,\ldots,t_n)$$

$$= (qt_1^2;q^2)_b(\frac{q}{t_1^2};q^2)_b \,_qf_n(a,b,k;t_1,\ldots,t_n).$$

Then we have the symmetries

(3.47)
$$_q\alpha_n^v(a,b,k;t_1,\ldots,t_{v-2},t_v,t_{v-1},t_{v+1},\ldots,t_n)$$
$$= \,_q\alpha_n^v(a,b,k;t_1,\ldots,t_n), \quad 2 \le v \le n,$$

(3.48)
$$_q\epsilon_n(a,b,k;t_1,\ldots,t_{n-1},\frac{1}{t_n}) = \,_q\epsilon_n(a,b,k;t_1,\ldots,t_n),$$

and

(3.49)
$$_qf_n(a,b,k;\frac{q}{t_1},t_2,\ldots,t_n) = \,_qf_n(a,b,k;t_1,\ldots,t_n).$$

Proof. We have

(3.50)
$$_q\alpha_n^v(a,b,k;t_1,\ldots,t_n) = \,_q\text{short}_n(a;t_1,\ldots,t_n) \,_q\text{long}_n(b;t_1,\ldots,t_n)$$
$$\times \,_q\text{ex}_n(k;t_1,\ldots,t_n) \,_qa_{n-1}^v(k;t_1,\ldots,t_n), \quad 2 \le v \le n.$$

By (3.33) and (3.43), we have that $_q\text{short}_n(a;t_1,\ldots,t_n)$, $_q\text{long}_n(b;t_1,\ldots,t_n)$ and $_q\text{ex}_n(k;t_1,\ldots,t_n)$ are symmetric in $t_1,\ldots,t_n$. The result (3.47) follows by the symmetry (3.24) of $_qa_{n-1}^v(k;t_1,\ldots,t_n)$. We have

(3.51)
$$_q\epsilon_n(a,b,k;t_1,\ldots,t_n)$$
$$= \prod_{i=1}^{n-1}(t_i)_a(\frac{q}{t_i})_a \,_q\text{long}_n(b;t_1,\ldots,t_n) \,_qd_n(k;t_1,\ldots,t_n),$$
$$_qf_n(a,b,k;t_1,\ldots,t_n)$$
$$= \,_q\text{short}_n(a;t_1,\ldots,t_n) \prod_{i=2}^{n}(qt_i^2;q^2)_b(\frac{q}{t_i^2};q^2)_b \,_qd_n(k;t_1,\ldots,t_n).$$

The results (3.48) and (3.49) follow by the symmetries (3.35) of $_qd_n(k;t_1,\ldots,t_n)$ and the symmetries (3.44) of $_q\text{short}_n(a;t_1,\ldots,t_n)$ and $_q\text{long}_n(b;t_1,\ldots,t_n)$. $\quad\square$

We may extend A_{n-1} to B_n, C_n or D_n by adding one positive simple root. By (3.24) or the geometry (3.15) of the simple roots of D_n, we have

$$(3.52) \qquad e_1 - e_v = w_v(e_{v-1} - e_v) \in \mathcal{B}(w_v(D_n^+)), \quad 2 \le v \le n.$$

By the geometry (3.15) of the simple roots of B_n, we have

$$(3.53) \qquad e_1 = w_n(e_n) \in \mathcal{B}(w_n(B_n^+)).$$

Set

$$(3.54) \qquad \tilde{w}_v = \sigma_v \cdots \sigma_{n-1} \sigma_{e_n} \sigma_{n-1} \cdots \sigma_1, \quad 1 \le v \le n.$$

Observe that

$$(3.55) \qquad \tilde{w}_v(e_1, \ldots, e_n) = (e_2, \ldots, e_v, -e_1, e_{v+1}, \ldots, e_n), \quad 1 \le v \le n.$$

By the geometry (3.15) of the simple roots of D_n, we have

$$(3.56) \qquad e_1 + e_v = \tilde{w}_v(e_{v-1} - e_v) \in \mathcal{B}(\tilde{w}_v(D_n^+)), \quad 2 \le v \le n.$$

By the geometry (3.15) of the simple roots of C_n, we have

$$(3.57) \qquad -2e_1 = \tilde{w}_1(2e_1) \in \mathcal{B}(\tilde{w}_1(C_n^+)).$$

The following lemma explicitly expresses the geometry of the simple roots of B_n and C_n in terms of $_qbc_n(a, b, k; t_1, \ldots, t_n)$.

Lemma 6.

$$_qbc_n(a, b, k; t_2, \ldots, t_{v-1}, t_1, t_v, \ldots, t_n)$$

$$(3.58)$$
$$= \prod_{i=1}^{n} (t_i)_a \left(\frac{q}{t_i}\right)_a (qt_i^2; q^2)_b \left(\frac{q}{t_i^2}; q^2\right)_b$$

$$\times \prod_{j=2}^{v-1} \left(q\frac{t_1}{t_j}\right)_k \left(\frac{t_j}{t_1}\right)_k (t_1 t_j)_k \left(\frac{q}{t_1 t_j}\right)_k \prod_{j=v}^{n} \left(\frac{t_1}{t_j}\right)_k \left(q\frac{t_j}{t_1}\right)_k (t_1 t_j)_k \left(\frac{q}{t_1 t_j}\right)_k$$

$$\times \prod_{2 \le i < j \le n} \left(\frac{t_i}{t_j}\right)_k \left(q\frac{t_j}{t_i}\right)_k (t_i t_j)_k \left(\frac{q}{t_i t_j}\right)_k, \quad 2 \le v \le n+1,$$

and

$$_qbc_n\left(a, b, k; t_2, \ldots, t_v, \frac{1}{t_1}, t_{v+1}, \ldots, t_n\right)$$

$$(3.59)$$
$$= (qt_1)_a \left(\frac{1}{t_1}\right)_a (qt_1^2; q^2)_b \left(\frac{q}{t_1^2}; q^2\right)_b \prod_{i=2}^{n} (t_i)_a \left(\frac{q}{t_i}\right)_a (qt_i^2; q^2)_b \left(\frac{q}{t_i^2}; q^2\right)_b$$

$$\times \prod_{j=2}^{v} \left(q\frac{t_1}{t_j}\right)_k \left(\frac{t_j}{t_1}\right)_k (t_1 t_j)_k \left(\frac{q}{t_1 t_j}\right)_k \prod_{j=v+1}^{n} \left(q\frac{t_1}{t_j}\right)_k \left(\frac{t_j}{t_1}\right)_k (qt_1 t_j)_k \left(\frac{1}{t_1 t_j}\right)_k$$

$$\times \prod_{2 \le i < j \le n} \left(\frac{t_i}{t_j}\right)_k \left(q\frac{t_j}{t_i}\right)_k (t_i t_j)_k \left(\frac{q}{t_i t_j}\right)_k, \quad 1 \le v \le n.$$

Proof. The results are easily verified by inspection.   □

Observe that (3.53) and (3.57) are given by the $v = n + 1$ case of (3.58) and the $v = 1$ case of (3.59), respectively.

4. The q-engine of our q-machine

In this section, we introduce the partial q-derivative and show that if $f(t_1, \ldots, t_n)$ has a Laurent expansion at $t_1 = 0$, then $\partial_q/\partial_q t_1 f(t_1, \ldots, t_n)$ has no residue at $t_1 = 0$. This q-analogue of (2.10) is the q-engine of our q-machine. It is equivalent to the fact that if $f(t_1, \ldots, t_n)$ has a Laurent expansion at $t_1 = 0$, then the constant term $[1]f(t_1, \ldots, t_n)$ is fixed by $t_1 \to qt_1$. Lemma 7 gives an identity for the partial q-derivative of ${}_q bc_n(a, b, k; t_1, \ldots, t_n)$ and shows that each term occurring therein is antisymmetric under a certain substitution. This is a q-analogue of (2.32). We apply the q-engine to $f(t_1, \ldots, t_n) = {}_q bc_{n,0,r}(a, b, k; t_1, \ldots, t_n)$, $r \geq 1$, and obtain the q-analogue Lemma 8 of (2.34).

Recall [Ka3] that the q-derivative

$$(4.1) \qquad \frac{d_q}{d_q t} f(t) = \frac{f(t) - f(qt)}{t(1 - q)}$$

satisfies the product rule

$$
\begin{aligned}
(4.2) \qquad \frac{d_q}{d_q t} \prod_{\nu=1}^{u} f_\nu(t) &= \frac{1}{t(1-q)} \left(\prod_{\nu=1}^{u} f_\nu(t) - \prod_{\nu=1}^{u} f_\nu(qt) \right) \\
&= \frac{1}{t(1-q)} \sum_{\nu=1}^{u} \prod_{i=1}^{\nu-1} f_i(qt) \left(f_\nu(t) - f_\nu(qt) \right) \prod_{j=\nu+1}^{u} f_j(t) \\
&= \sum_{\nu=1}^{u} \prod_{i=1}^{\nu-1} f_i(qt) \left(\frac{d_q}{d_q t} f_\nu(t) \right) \prod_{j=\nu+1}^{u} f_j(t).
\end{aligned}
$$

Using the obvious notation for partial q-derivatives, we have

$$(4.3) \qquad t_1 \frac{\partial_q}{\partial_q t_1} f(t_1, \ldots, t_n) = \frac{1}{(1-q)} \left(f(t_1, \ldots, t_n) - f(qt_1, t_2, \ldots, t_n) \right).$$

If $f(t_1, \ldots, t_n)$ has a Laurent expansion at $t_1 = 0$, then the constant term $[1]$ $f(t_1, \ldots, t_n)$ is fixed by $t_1 \to qt_1$. Hence we have the equivalent conclusion that

$$(4.4) \qquad [1] t_1 \frac{\partial_q}{\partial_q t_1} f(t_1, \ldots, t_n) = 0,$$

which is the q-engine of our q-machine. This is a q-analogue of (2.10).

A simple calculation gives

$$
\begin{aligned}
(4.5) \qquad (q\frac{s}{t})_k (\frac{t}{s})_k &= \frac{(1 - q^k \frac{s}{t})}{(1 - \frac{s}{t})} (\frac{s}{t})_k \frac{(1 - \frac{t}{s})}{(1 - q^k \frac{t}{s})} (q\frac{t}{s})_k \\
&= -\frac{t}{s} \frac{(1 - q^k \frac{s}{t})}{(1 - q^k \frac{t}{s})} (\frac{s}{t})_k (q\frac{t}{s})_k \\
&= \frac{(q^k s - t)}{(s - q^k t)} (\frac{s}{t})_k (q\frac{t}{s})_k.
\end{aligned}
$$

We obtain

$$
s \frac{\partial_q}{\partial_q s} \left((\frac{s}{t})_k (q\frac{t}{s})_k \right) = \frac{1}{(1-q)} \left((\frac{s}{t})_k (q\frac{t}{s})_k - (q\frac{s}{t})_k (\frac{t}{s})_k \right)
$$

(4.6)
$$
= \frac{1}{(1-q)} \left(1 - \frac{(q^k s - t)}{(s - q^k t)} \right) (\frac{s}{t})_k (q\frac{t}{s})_k
$$

$$
= \frac{(1-q^k)}{(1-q)} \frac{(s+t)}{(s - q^k t)} (\frac{s}{t})_k (q\frac{t}{s})_k.
$$

Let v run from 2 to n and use (4.6) to q-differentiate the factor $(t_1/t_v)_k (qt_v/t_1)_k$ of $_q a_{n-1}(k; t_1, \ldots, t_n)$. We set $s = t_1$ and $t = t_v$, $2 \le v \le n$, in (4.6) and use (3.28) to effect the substitution $t_1 \to qt_1$ in each factor that is q-differentiated. By the product rule (4.2) for q-derivatives, we obtain

(4.7)
$$
t_1 \frac{\partial_q}{\partial_q t_1} \left({_q a_{n-1}(k; t_1, \ldots, t_n)} \right)
$$
$$
= \frac{(1-q^k)}{(1-q)} \sum_{v=2}^{n} \frac{(t_1 + t_v)}{(t_1 - q^k t_v)} \, {_q a_{n-1}(k; t_2, \ldots, t_{v-1}, t_1, t_v, \ldots, t_n)}.
$$

Let $s \partial_q / \partial_q s \left((s/t)_k (qt/s)_k \right)$ be denoted by $_q h(k; s, t)$. By (4.6) we see that $_q h(k; s, t)$ is antisymmetric under $s \leftrightarrow t$. Thus

(4.8)
$$
_q h(k; s, t) = \frac{(1-q^k)}{(1-q)} \frac{(s+t)}{(s - q^k t)} (\frac{s}{t})_k (q\frac{t}{s})_k
$$
$$
= - _q h(k; t, s).
$$

We may write (4.7) as

(4.9)
$$
t_1 \frac{\partial_q}{\partial_q t_1} \left({_q a_{n-1}(k; t_1, \ldots, t_n)} \right)
$$
$$
= \sum_{v=2}^{n} {_q h(k; t_1, t_v)} \, {_q a_{n-1}^{v}(k; t_2, \ldots, t_{v-1}, t_1, t_v, \ldots, t_n)}.
$$

Using the antisymmetry (4.8) of $_q h(k; s, t)$ and the symmetry (3.24) of $_q \alpha_{n-1}^{v}(k; t_1, \ldots, t_n)$, we see that the term occurring in the sum on the right side of (4.7) is antisymmetric under

(4.10) $t_1 \leftrightarrow t_v, \ 2 \le v \le n.$

Observe that we may continue the q-differentiation in (4.7) to obtain

(4.11)

$$t_1 \frac{\partial_q}{\partial_q t_1} \left(f(t_1, \ldots, t_n) \, _q a_{n-1}(k; t_1, \ldots, t_n) \right)$$

$$= f(t_1, \ldots, t_n) \frac{(1-q^k)}{(1-q)} \sum_{v=2}^{n} \frac{(t_1 + t_v)}{(t_1 - q^k t_v)} \, _q a_{n-1}(k; t_2, \ldots, t_{v-1}, t_1, t_v, \ldots, t_n)$$

$$+ t_1 \frac{\partial_q}{\partial_q t_1} \left(f(t_1, \ldots, t_n) \right) \, _q a_{n-1}(k; q t_1, t_2, \ldots, t_n).$$

Setting $t = 1$ and replacing k by a in (4.6) yields

(4.12)
$$s \frac{\partial_q}{\partial_q s} \left((s)_a (\frac{q}{s})_a \right) = \, _q h(a; s, 1)$$
$$= \frac{(1-q^a)}{(1-q)} \frac{(1+s)}{(s-q^a)} (s)_a (\frac{q}{s})_a.$$

Replacing t by $1/t$ in (4.6) and multiplying numerator and denominator by t, we have

(4.13)
$$s \frac{\partial_q}{\partial_q s} \left((st)_k (\frac{q}{st})_k \right) = \, _q h(k; s, \frac{1}{t})$$
$$= \frac{(1-q^k)}{(1-q)} \frac{(st+1)}{(st-q^k)} (st)_k (\frac{q}{st})_k.$$

By the definition (4.1) of the partial q-derivative, we have

(4.14)
$$s \frac{\partial_q}{\partial_q s} \left((qs^2; q^2)_b (\frac{q}{s^2}; q^2)_b \right)$$
$$= \frac{1}{(1-q)} \left((qs^2; q^2)_b (\frac{q}{s^2}; q^2)_b - (q^3 s^2; q^2)_b (\frac{1}{qs^2}; q^2)_b \right)$$
$$= \frac{(1-q^2)}{(1-q)} \, _{q^2} h(b; qs^2, 1)$$
$$= \frac{(1-q^{2b})}{(1-q)} \frac{(qs^2+1)}{(qs^2 - q^{2b})} (qs^2; q^2)_b (\frac{q}{s^2}; q^2)_b.$$

The following lemma is a q-analogue of (2.32).

Lemma 7.

$$
t_1 \frac{\partial_q}{\partial_q t_1} \left({}_qbc_n(a,b,k;t_1,\dots,t_n) \right)
$$

$$
= \frac{(1-q^k)}{(1-q)} \sum_{v=2}^{n} \frac{(t_1+t_v)}{(t_1-q^k t_v)} \, {}_qbc_n(a,b,k;t_2,\dots,t_{v-1},t_1,t_v,\dots,t_n)
$$

(4.15)
$$
+ \frac{(1-q^a)}{(1-q)} \frac{(t_1+1)}{(t_1-q^a)} \, {}_qbc_n(a,b,k;t_2,\dots,t_n,t_1)
$$

$$
+ \frac{(1-q^k)}{(1-q)} \sum_{v=2}^{n} \frac{(t_1 t_v+1)}{(t_1 t_v-q^k)} \, {}_qbc_n(a,b,k;t_2,\dots,t_v,\frac{1}{t_1},t_{v+1},\dots,t_n)
$$

$$
+ \frac{(1-q^{2b})}{(1-q)} \frac{(qt_1^2+1)}{(qt_1^2-q^{2b})} \, {}_qbc_n(a,b,k;\frac{1}{t_1},t_2,\dots,t_n)
$$

and the terms occurring on the right side of (4.15) are antisymmetric under the substitutions

(4.16)
$$
t_1 \leftrightarrow t_v, \ 2 \le v \le n, \quad t_1 \leftrightarrow \frac{1}{t_1},
$$
$$
t_1 \leftrightarrow \frac{1}{t_v}, \ 2 \le v \le n, \quad t_1 \leftrightarrow \frac{1}{qt_1},
$$

respectively.

Proof. We q-differentiate all of the factors of ${}_qbc_n(a,b,k;t_1,\dots,t_n)$ which depend on t_1 in a certain order and use the product rule (4.2) for q-derivatives. Letting v run from 2 to n, we set $s=t_1$ and $t=t_v$ in (4.6) and q-differentiate the factor $(t_1/t_v)_k(qt_v/t_1)_k$. By Lemma 6 (3.58), we may effect the substitution $t_1 \to qt_1$ in each factor that is q-differentiated. Setting $s=t_1$ in (4.12), we q-differentiate the factor $(t_1)_a(q/t_1)_a$. Letting v run from n down to 2, we set $s=t_1$ and $t=t_v$ in (4.13) and q-differentiate the factor $(t_1 t_v)_k(q/t_1 t_v)_k$. By Lemma 6 (3.59), we may effect the substitution $t_1 \to qt_1$ in each factor that is q-differentiated. Setting $s=t_1$ in (4.14), we q-differentiate the factor $(qt_1^2;q^2)_b(q/t_1^2;q^2)_b$. The partial q-derivative identity (4.15) then follows by the product rule (4.2) for q-derivatives.

We may write (4.15) as

$$t_1 \frac{\partial_q}{\partial_q t_1} \left({}_q bc_n(a,b,k;t_1,\ldots,t_n) \right)$$

$$= \sum_{v=2}^{n} {}_q h(k;t_1,t_v) \, {}_q \alpha_n^v(a,b,k;t_2,\ldots,t_{v-1},t_1,t_v,\ldots,t_n)$$

(4.17)
$$+ \, {}_q h(a;t_1,1) \, {}_q \epsilon_n(a,b,k;t_2,\ldots,t_n,t_1)$$

$$+ \sum_{v=2}^{n} {}_q h(k;t_1,\frac{1}{t_v}) \, {}_q \alpha_n^v(a,b,k;t_2,\ldots,t_v,\frac{1}{t_1},t_{v+1},\ldots,t_n)$$

$$+ \frac{(1-q^{2b})}{(1-q^b)} \, {}_{q^2} h(b;qt_1^2,1) \, {}_q f_n(a,b,k;\frac{1}{t_1},t_2,\ldots,t_n).$$

By Lemma 5 (3.47)–(3.49), the functions ${}_q\alpha_n^v(a,b,k;t_2,\ldots,t_{v-1},t_1,t_v,\ldots,t_n)$, ${}_q\epsilon_n(a,b,k;t_2,\ldots,t_n,t_1)$, ${}_q\alpha_n^v(a,b,k;t_2,\ldots,t_v,1/t_1,t_{v+1},\ldots,t_n)$ and ${}_qf_n(a,b,k;$ $1/t_1,t_2,\ldots,t_n)$ occurring on the right side of (4.17) are symmetric under the corresponding substitutions (4.16). The first and third antisymmetries of (4.16) follow by setting $s=t_1$, $t=t_v$, and $s=t_1$, $t=1/t_1$, respectively, in (4.8). By (4.6) we see that ${}_q h(k;s,t)$ is antisymmetric

(4.18)
$$\quad {}_q h(k;s,t) = -\,{}_q h(k;\frac{1}{s},\frac{1}{t}).$$

under $s \leftrightarrow 1/s$, $t \leftrightarrow 1/t$. The second and fourth antisymmetries of (4.16) follow by taking $s=t_1$ and $s=qt_1^2$ in (4.18), respectively. Observe that the last antisymmetry of (4.16) also follows by inspection of the first expression on the right side of (4.14). $\quad \square$

If we replace the arguments in the terms occurring on the right side of (4.15) by $(t_1,\ldots,t_n)$, then the substitutions in (4.16) become $t_{v-1} \leftrightarrow t_v$, $2 \le v \le n$, $t_n \leftrightarrow 1/t_n$ and $t_1 \leftrightarrow q/t_1$, which correspond to the simple reflections of B_n and C_n. We prefer to do this latter.

Fix $r \ge 1$ and apply the q-engine (4.4) to $f(t_1,\ldots,t_n) = {}_q bc_{n,0,r}(a,b,k;t_1,\ldots,t_n)$. We have

(4.19)
$$[1]\, t_1 \frac{\partial_q}{\partial_q t_1} \left(\prod_{i=1}^{r} t_i \, {}_q bc_n(a,b,k;t_1,\ldots,t_n) \right) = 0, \quad r \ge 1.$$

Since $r \ge 1$, we see that t_1 is a factor of $\prod_{i=1}^{r} t_i$. It is q-differentiated by

(4.20)
$$s \frac{\partial_q}{\partial_q s} s = s \frac{(s-qs)}{s(1-q)} = s.$$

By the product rule (4.2) for q-derivatives, we have

(4.21)
$$t_1 \frac{\partial_q}{\partial_q t_1} \left(\prod_{i=1}^r t_i \, _q bc_n(a,b,k;t_1,\ldots,t_n) \right)$$
$$= \prod_{i=1}^r t_i \, t_1 \frac{\partial_q}{\partial_q t_1} \left(\, _q bc_n(a,b,k;t_1,\ldots,t_n) \right)$$
$$+ \prod_{i=1}^r t_i \, _q bc_n(a,b,k;qt_1,t_2,\ldots,t_n), \quad r \geq 1.$$

Substituting the partial q-derivative identity (4.15) into (4.21) gives
(4.22)
$$t_1 \frac{\partial_q}{\partial_q t_1} \left(\prod_{i=1}^r t_i \, _q bc_n(a,b,k;t_1,\ldots,t_n) \right)$$
$$= \frac{(1-q^k)}{(1-q)} \sum_{v=2}^n \frac{(t_1+t_v)}{(t_1-q^k t_v)} \prod_{i=1}^r t_i \, _q bc_n(a,b,k;t_2,\ldots,t_{v-1},t_1,t_v,\ldots,t_n)$$
$$+ \frac{(1-q^a)}{(1-q)} \frac{(t_1+1)}{(t_1-q^a)} \prod_{i=1}^r t_i \, _q bc_n(a,b,k;t_2,\ldots,t_n,t_1)$$
$$+ \frac{(1-q^k)}{(1-q)} \sum_{v=2}^n \frac{(t_1 t_v+1)}{(t_1 t_v-q^k)} \prod_{i=1}^r t_i \, _q bc_n(a,b,k;t_2,\ldots,t_v,\frac{1}{t_1},t_{v+1},\ldots,t_n)$$
$$+ \frac{(1-q^{2b})}{(1-q)} \frac{(qt_1^2+1)}{(qt_1^2-q^{2b})} \prod_{i=1}^r t_i \, _q bc_n(a,b,k;\frac{1}{t_1},t_2,\ldots,t_n)$$
$$+ \prod_{i=1}^r t_i \, _q bc_n(a,b,k;qt_1,t_2,\ldots,t_n), \quad r \geq 1.$$

The following lemma is a q-analogue of (2.34).

Lemma 8.

(4.23)
$$0 = (1-q^k) \sum_{v=2}^n {}_q A_{n,0,r}^v(a,b,k) + (1-q^a) \, _q E_{n,0,r}(a,b,k)$$
$$+ (1-q^k) \sum_{v=2}^n {}_q K_{n,0,r}^v(a,b,k) + (1-q^{2b}) \, _q F_{n,0,r}(a,b,k)$$
$$+ (1-q) \, _q Z_{n,0,r}(a,b,k), \quad r \geq 1,$$

where

$$_qA^v_{n,0,r}(a,b,k) = [1]\frac{(t_1+t_v)}{(t_1-q^kt_v)}\prod_{i=1}^{r}t_i \; _qbc_n(a,b,k;t_2,\ldots,t_{v-1},t_1,t_v,\ldots,t_n),$$

(4.24) $2 \le v \le n,$

(4.25)

$$_qE_{n,0,r}(a,b,k) = [1]\frac{(t_1+1)}{(t_1-q^a)}\prod_{i=1}^{r}t_i \; _qbc_n(a,b,k;t_2,\ldots,t_n,t_1),$$

$$_qK^v_{n,0,r}(a,b,k) = [1]\frac{(t_1t_v+1)}{(t_1t_v-q^k)}\prod_{i=1}^{r}t_i \; _qbc_n(a,b,k;t_2,\ldots,t_v,\frac{1}{t_1},t_{v+1},\ldots,t_n),$$

(4.26) $2 \le v \le n,$

(4.27)

$$_qF_{n,0,r}(a,b,k) = [1]\frac{(qt_1^2+1)}{(qt_1^2-q^{2b})}\prod_{i=1}^{r}t_i \; _qbc_n(a,b,k;\frac{1}{t_1},t_2,\ldots,t_n),$$

(4.28)

$$_qZ_{n,0,r}(a,b,k) = [1]\prod_{i=1}^{r}t_i \; _qbc_n(a,b,k;qt_1,t_2,\ldots,t_n).$$

Proof. By the q-engine (4.4), the constant term in (4.22) is 0. The result (4.23) follows by extracting the constant term of (4.22) and multiplying by $(1-q)$. $\square$

5. Removing the denominators

In this section, we use the antisymmetries of Lemma 7 (4.16) to remove the factors occurring in the denominators of Lemma 8 (4.24)–(4.27). We obtain Lemma 9 which gives q-analogues of (2.40)–(2.45).

The following lemma removes the factors occurring in the denominators of Lemma 8 (4.24)–(4.27), giving q-analogues of (2.40)–(2.45).

Lemma 9. *Let* $r \geq 1$. *Then*

(5.1)
$$_q A_{n,0,r}^v(a,b,k) = 0, \quad 2 \leq v \leq r,$$

$$_q A_{n,0,r}^v(a,b,k) = \frac{1}{(1+q^k)} [1] (t_1 + t_v) \prod_{i=2}^{r} t_i$$

(5.2)
$$\times \; _q bc_n(a,b,k;t_2,\dots,t_{v-1},t_1,t_v,\dots,t_n), \quad r < v \leq n,$$

(5.3)
$$_q E_{n,0,r}(a,b,k) = \frac{1}{(1+q^{2a})} [1] (t_1 + 1)(1 + \frac{q^a}{t_1}) \prod_{i=2}^{r} t_i \; _q bc_n(a,b,k;t_2,\dots,t_n,t_1),$$

$$_q K_{n,0,r}^v(a,b,k) = \frac{1}{(1+q^{2k})} [1] (t_1 t_v + 1)(1 + \frac{q^k}{t_1 t_v}) \prod_{\substack{i=2 \\ i \neq v}}^{r} t_i$$

(5.4)
$$\times \; _q bc_n(a,b,k;t_2,\dots,t_v,\frac{1}{t_1},t_{v+1},\dots,t_n), \quad 2 \leq v \leq r,$$

$$_q K_{n,0,r}^v(a,b,k) = \frac{1}{(1+q^k)} [1] (t_1 t_v + 1) \frac{1}{t_v} \prod_{i=2}^{r} t_i$$

(5.5)
$$\times \; _q bc_n(a,b,k;t_2,\dots,t_v,\frac{1}{t_1},t_{v+1},\dots,t_n), \quad r < v \leq n,$$

(5.6)
$$_q F_{n,0,r}(a,b,k) = \frac{1}{(1+q^{2b})} [1] (qt_1^2 + 1) \frac{1}{qt_1} \prod_{i=2}^{r} t_i \; _q bc_n(a,b,k;\frac{1}{t_1},t_2,\dots,t_n),$$

Proof. The constant term in a polynomial is fixed by a substitution. Making the substitution $t_1 \leftrightarrow t_v$, $2 \leq v \leq n$, in Lemma 8 (4.24) and using the first antisymmetry of Lemma 7 (4.16), we obtain

$$_q A_{n,0,r}^v(a,b,k)$$

(5.7)
$$= -[1] \frac{(t_1 + t_v)}{(t_1 - q^k t_v)} \prod_{i=1}^{r} t_i \; _q bc_n(a,b,k;t_2,\dots,t_{v-1},t_1,t_v,\dots,t_n),$$
$$2 \leq v \leq r,$$

32

and

$$_q A_{n,0,r}^v(a,b,k)$$

(5.8)
$$= -[1] \frac{(t_1 + t_v)}{(t_1 - q^k t_v)} t_v \prod_{i=2}^{r} t_i \, _q bc_n(a,b,k;t_2,\dots,t_{v-1},t_1,t_v,\dots,t_n),$$
$$r < v \le n.$$

Comparing (4.24) and (5.7), we see that the result (5.1) holds. Adding q^k times (5.8) to (4.24) yields
(5.9)

$$(1 + q^k) \, _q A_{n,0,r}^v(a,b,k)$$

$$= [1] \frac{(t_1 + t_v)}{(t_1 - q^k t_v)} (t_1 - q^k t_v) \prod_{i=2}^{r} t_i \, _q bc_n(a,b,k;t_2,\dots,t_{v-1},t_1,t_v,\dots,t_n)$$

$$= [1](t_1 + t_v) \prod_{i=2}^{r} t_i \, _q bc_n(a,b,k;t_2,\dots,t_{v-1},t_1,t_v,\dots,t_n), \quad r < v \le n.$$

Dividing (5.9) by $(1 + q^k)$, we obtain the result (5.2).

Making the substitution $t_1 \leftrightarrow 1/t_1$ in Lemma 8 (4.25) and using the second antisymmetry of Lemma 7 (4.16), we obtain

(5.10) $$_q E_{n,0,r}(a,b,k) = -[1] \frac{(t_1 + 1)}{(t_1 - q^a)} \frac{1}{t_1} \prod_{i=2}^{r} t_i \, _q bc_n(a,b,k;t_2,\dots,t_n,t_1).$$

Observe that

(5.11)
$$\frac{(t_1 - \frac{q^{2a}}{t_1})}{(t_1 - q^a)} = \frac{1}{t_1}(t_1 + q^a) = (1 + \frac{q^a}{t_1}).$$

Adding q^{2a} times (5.10) to (4.25) and simplifying by (5.11), we obtain

$$(1 + q^{2a}) \, _q E_{n,0,r}(a,b,k)$$

(5.12)
$$= [1] \frac{(t_1 + 1)}{(t_1 - q^a)} (t_1 - \frac{q^{2a}}{t_1}) \prod_{i=2}^{r} t_i \, _q bc_n(a,b,k;t_2,\dots,t_n,t_1)$$

$$= [1](t_1 + 1)(1 + \frac{q^a}{t_1}) \prod_{i=2}^{r} t_i \, _q bc_n(a,b,k;t_2,\dots,t_n,t_1).$$

Dividing (5.12) by $(1 + q^{2a})$, we obtain the result (5.3).

Making the substitution $t_1 \leftrightarrow 1/t_v$, $2 \le v \le n$, in Lemma 8 (4.26) and using the third antisymmetry of Lemma 7 (4.16), we obtain
(5.13)
$$_q K_{n,0,r}^v(a,b,k)$$

$$= -[1] \frac{(t_1 t_v + 1)}{(t_1 t_v - q^k)} \frac{1}{t_1 t_v} \prod_{\substack{i=2 \\ i \ne v}}^{r} t_i \, _q bc_n(a,b,k;t_2,\dots,t_v,\frac{1}{t_1},t_{v+1},\dots,t_n),$$
$$2 \le v \le r,$$

and

$$_qK_{n,0,r}^v(a,b,k)$$

(5.14)
$$= -[1]\frac{(t_1t_v+1)}{(t_1t_v-q^k)}\frac{1}{t_v}\prod_{i=2}^r t_i \; _qbc_n(a,b,k;t_2,\ldots,t_v,\frac{1}{t_1},t_{v+1},\ldots,t_n),$$
$$r < v \le n.$$

Observe that

(5.15)
$$\frac{(t_1t_v-\dfrac{q^{2k}}{t_1t_v})}{(t_1t_v-q^k)}=\frac{1}{t_1t_v}(t_1t_v+q^{2k})=(1+\frac{q^k}{t_1t_v}).$$

Adding q^{2k} times (5.13) to (4.26) and simplifying by (5.15), we obtain
(5.16)

$$(1+q^{2k})\;_qK_{n,0,r}^v(a,b,k)$$

$$= [1]\frac{(t_1t_v+1)}{(t_1t_v-q^k)}(t_1t_v-\frac{q^{2k}}{t_1t_v})\prod_{\substack{i=2\\i\ne v}}^r t_i\;_qbc_n(a,b,k;t_2,\ldots,t_v,\frac{1}{t_1},t_{v+1},\ldots,t_n)$$

$$= [1](t_1t_v+1)(1+\frac{q^k}{t_1t_v})\prod_{\substack{i=2\\i\ne v}}^r t_i\;_qbc_n(a,b,k;t_2,\ldots,t_v,\frac{1}{t_1},t_{v+1},\ldots,t_n),$$
$$2 \le v \le r.$$

Dividing (5.16) by $(1+q^{2k})$, we obtain the result (5.4). Observe that

(5.17)
$$\frac{(t_1-\dfrac{q^k}{t_v})}{(t_1t_v-q^k)}=\frac{1}{t_v}.$$

Adding q^k times (5.14) to (4.26) and simplifying by (5.17), we obtain

(5.18) $$(1+q^k)\;_qK_{n,0,r}^v(a,b,k)$$

$$= [1]\frac{(t_1t_v+1)}{(t_1t_v-q^k)}(t_1-\frac{q^k}{t_v})\prod_{i=2}^r t_i\;_qbc_n(a,b,k;t_2,\ldots,t_v,\frac{1}{t_1},t_{v+1},\ldots,t_n)$$

$$= [1](t_1t_v+1)\frac{1}{t_v}\prod_{i=2}^r t_i\;_qbc_n(a,b,k;t_2,\ldots,t_v,\frac{1}{t_1},t_{v+1},\ldots,t_n), \quad r < v \le n.$$

Dividing (5.18) by $(1+q^k)$, we obtain the result (5.5).

Making the substitution $t_1 \leftrightarrow 1/qt_1$ in Lemma 8 (4.27) and using the fourth antisymmetry of Lemma 7 (4.16), we obtain

(5.19) $$_qF_{n,0,r}(a,b,k) = -[1]\frac{(qt_1^2+1)}{(qt_1^2-q^{2b})}\frac{1}{qt_1}\prod_{i=2}^r t_i\;_qbc_n(a,b,k;\frac{1}{t_1},t_2,\ldots,t_n),$$

Observe that

$$(5.20) \qquad \frac{(t_1 - \frac{q^{2b}}{qt_1})}{(qt_1^2 - q^{2b})} = \frac{1}{qt_1}.$$

Adding q^{2b} times (4.27) to (5.19) and simplifying by (5.20), we obtain

$$(5.21) \qquad \begin{aligned} &(1 + q^{2b})\, {}_qF_{n,0,r}(a,b,k) \\ &\quad = [1]\,\frac{(qt_1^2 + 1)}{(qt_1^2 - q^{2b})}\,(t_1 - \frac{q^{2b}}{qt_1})\prod_{i=2}^{r} t_i \; {}_qbc_n(a,b,k;\frac{1}{t_1},t_2,\dots,t_n) \\ &\quad = [1]\,(qt_1^2 + 1)\,\frac{1}{qt_1}\,\prod_{i=2}^{r} t_i \; {}_qbc_n(a,b,k;\frac{1}{t_1},t_2,\dots,t_n). \end{aligned}$$

Dividing (5.21) by $(1 + q^{2b})$, we obtain the result (5.6). $\quad\square$

6. The q-transportation theory for BC_n

In this section, we use the basic property of the simple reflections of B_n and C_n to give a local proof of the q-transportation theory for BC_n. The global version of the integral formulation of the q-transportation theory for A_{n-1} is the q-analogue [Ka2, (4.14)] of (2.5). This was used in the proof [Ka3] of Askey's q-Selberg integral. Lemma 10 is the local version of the constant term formulation. Lemma 11, which we call the q-transportation theory for A_{n-1}, explicitly expresses Lemma 10 in terms of $_qbc_n(a, b, k; t_1, \dots, t_n)$. Lemmas 12 and 14 extend the q-transportation theory for A_{n-1} to B_n and C_n, respectively. Lemmas 13 and 15 explicitly express the results in terms of $_qbc_n(a, b, k; t_1, \dots, t_n)$. Lemmas 11, 13 and 15, which give q-analogues of the transportation property (2.14), constitute the q-transportation theory for BC_n.

The global version of the integral formulation of the q-transportation theory for A_{n-1} is the q-analogue [Ka2, (4.14)] of (2.5). This was proved [Ka2] by extending the Vandermonde determinant (1.8) and used in the proof [Ka3] of Askey's q-Selberg integral. The following lemma is the local version of the constant term formulation of the q-transportation theory for A_{n-1}.

Lemma 10. *Let* $\mathrm{Sym}(s, t)$ *be symmetric in* s *and* t *and have a Laurent expansion at* $s = t = 0$. *Then*

$$(6.1) \qquad [1]\, t\, (1 - \frac{s}{t})(1 - Q\frac{t}{s})\,\mathrm{Sym}(s, t) = Q\,[1]\, s\, (1 - \frac{s}{t})(1 - Q\frac{t}{s})\,\mathrm{Sym}(s, t),$$

$$(6.2) \qquad [1]\, \frac{1}{t}\, (1 - \frac{s}{t})(1 - Q\frac{t}{s})\,\mathrm{Sym}(s, t) = \frac{1}{Q}\,[1]\, \frac{1}{s}\, (1 - \frac{s}{t})(1 - Q\frac{t}{s})\,\mathrm{Sym}(s, t),$$

and

$$(6.3) \qquad \begin{aligned}
&[1]\, \frac{s}{t}\, (1 - \frac{s}{t})(1 - Q\frac{t}{s})\,\mathrm{Sym}(s, t) \\
&\qquad = (\frac{1}{Q} - 1)\,[1]\, (1 - \frac{s}{t})(1 - Q\frac{t}{s})\,\mathrm{Sym}(s, t) \\
&\qquad\quad + \frac{1}{Q}\,[1]\, \frac{t}{s}\, (1 - \frac{s}{t})(1 - Q\frac{t}{s})\,\mathrm{Sym}(s, t).
\end{aligned}$$

Proof. Since $t - s$ changes sign under $s \leftrightarrow t$, we have

$$(6.4) \qquad\qquad\qquad [1]\, (t - s)\,\mathrm{Sym}(s, t) = 0.$$

Thus the left side of (6.1) becomes

$$(6.5) \qquad \begin{aligned}
&[1]\, t\, (1 - \frac{s}{t})(1 - Q\frac{t}{s})\,\mathrm{Sym}(s, t) \\
&\qquad = [1]\, (t - s)\,\mathrm{Sym}(s, t) - Q\,[1]\, \frac{t}{s}\, (t - s)\,\mathrm{Sym}(s, t) \\
&\qquad = -Q\,[1]\, \frac{t}{s}\, (t - s)\,\mathrm{Sym}(s, t).
\end{aligned}$$

Interchanging s and t, we have

$$(6.6) \qquad [1]\, t\,(1 - \frac{s}{t})(1 - Q\frac{t}{s})\, \mathrm{Sym}(s,t) = Q\,[1]\,\frac{s}{t}\,(t-s)\,\mathrm{Sym}(s,t).$$

Similarly, the constant term on the right side of (6.1) is

$$[1]\, s\,(1 - \frac{s}{t})(1 - Q\frac{t}{s})\, \mathrm{Sym}(s,t)$$

$$(6.7) \qquad = [1]\, s\,(1 - \frac{s}{t})\, \mathrm{Sym}(s,t) - Q\,[1]\,(t-s)\,\mathrm{Sym}(s,t)$$

$$= [1]\,\frac{s}{t}\,(t-s)\,\mathrm{Sym}(s,t).$$

The result (6.1) follows by comparing (6.6) and (6.7). Using

$$(6.8) \qquad \frac{1}{t} = s\,\frac{1}{st} \quad \text{and} \quad \frac{1}{s} = t\,\frac{1}{st},$$

we see that the result (6.2) follows from (6.1). Incorporating $s+t$ into $\mathrm{Sym}(s,t)$, we see that (6.2) gives
$$(6.9)$$
$$[1]\,\frac{1}{t}\,(s+t)\,(1 - \frac{s}{t})(1 - Q\frac{t}{s})\, \mathrm{Sym}(s,t) = \frac{1}{Q}\,[1]\,\frac{1}{s}\,(s+t)\,(1 - \frac{s}{t})(1 - Q\frac{t}{s})\, \mathrm{Sym}(s,t).$$

The result (6.3) follows by subtracting

$$(6.10) \qquad [1]\,(1 - \frac{s}{t})(1 - Q\frac{t}{s})\, \mathrm{Sym}(s,t)$$

from both sides of (6.9). $\square$

    The following lemma, which we call the q-transportation theory for A_{n-1}, explicitly expresses Lemma 10 in terms of $_qbc_n(a,b,k;t_1,\ldots,t_n)$.

Lemma 11. *Let* $2 \le v \le n$. *If* $\omega(t_1,\ldots,t_n)$ *is symmetric*

$$(6.11) \qquad \omega(t_1,\ldots,t_n) = \omega(t_1,\ldots,t_{v-2},t_v,t_{v-1},t_{v+1},\ldots,t_n)$$

in t_{v-1} *and* t_v, *then*

$$[1]\,t_v\,\omega(t_1,\ldots,t_n)\, {}_qbc_n(a,b,k;t_1,\ldots,t_n)$$

$$(6.12)$$

$$= q^k\,[1]\,t_{v-1}\,\omega(t_1,\ldots,t_n)\, {}_qbc_n(a,b,k;t_1,\ldots,t_n),$$

$$[1]\,\frac{1}{t_v}\,\omega(t_1,\ldots,t_n)\, {}_qbc_n(a,b,k;t_1,\ldots,t_n)$$

$$(6.13)$$

$$= q^{-k}\,[1]\,\frac{1}{t_{v-1}}\,\omega(t_1,\ldots,t_n)\, {}_qbc_n(a,b,k;t_1,\ldots,t_n),$$

and

$$(6.14) \quad \begin{aligned} &[1]\frac{t_{v-1}}{t_v}\,\omega(t_1,\dots,t_n)\,{}_qbc_n(a,b,k;t_1,\dots,t_n)\\ &\quad = (q^{-k}-1)\,[1]\,\omega(t_1,\dots,t_n)\,{}_qbc_n(a,b,k;t_1,\dots,t_n)\\ &\quad\quad + q^{-k}\,[1]\,\frac{t_v}{t_{v-1}}\,\omega(t_1,\dots,t_n)\,{}_qbc_n(a,b,k;t_1,\dots,t_n). \end{aligned}$$

Proof. Observe that

$$(6.15) \quad (\frac{t_{v-1}}{t_v})_k(q\frac{t_v}{t_{v-1}})_k = (1-\frac{t_{v-1}}{t_v})(1-q^k\frac{t_v}{t_{v-1}})(q\frac{t_{v-1}}{t_v})_{k-1}(q\frac{t_v}{t_{v-1}})_{k-1}.$$

By the definition (3.46) of ${}_q\alpha_n^v(a,b,k;t_1,\dots,t_n)$, we have

$$(6.16) \quad \omega(t_1,\dots,t_n)\,{}_qbc_n(a,b,k;t_1,\dots,t_n) = (1-\frac{t_{v-1}}{t_v})(1-q^k\frac{t_v}{t_{v-1}})\,\mathrm{Sym}(t_{v-1},t_v),$$

where
$$(6.17)$$
$$\mathrm{Sym}(t_{v-1},t_v) = \omega(t_1,\dots,t_n)\,(q\frac{t_{v-1}}{t_v})_{k-1}(q\frac{t_v}{t_{v-1}})_{k-1}\,{}_q\alpha_n^v(a,b,k;t_1,\dots,t_n).$$

Using the symmetry Lemma 5 (3.47) of ${}_q\alpha_n^v(a,b,k;t_1,\dots,t_n)$, we see that $\mathrm{Sym}(t_{v-1}, t_v)$ is symmetric in t_{v-1} and t_v. Thus we may apply Lemma 10 with $\mathrm{Sym}(t_{v-1},t_v)$ given by (6.17) and $s = t_{v-1}$, $t = t_v$, $Q = q^k$. Observe that (6.1), (6.2) and (6.3) then become the results (6.12), (6.13) and (6.14), respectively. $\square$

The following lemma extends Lemma 10 from A_{n-1} to B_n.

Lemma 12. *Let*

$$(6.18) \quad \mathrm{T}(t) = \mathrm{T}(\frac{1}{t})$$

be invariant under $t \leftrightarrow 1/t$ *and have a Laurent expansion at* $t = 0$. *Then*

$$(6.19) \quad [1]\,(1+\frac{1}{t})\,(1-t)(1-\frac{Q}{t})\,\mathrm{T}(t) = Q\,[1]\,(1+t)\,(1-t)(1-\frac{Q}{t})\,\mathrm{T}(t).$$

Proof. Since

$$(6.20) \quad (1+\frac{1}{t})(1-t) = \frac{1}{t} - t$$

changes sign under $t \leftrightarrow 1/t$, we have

$$(6.21) \quad [1]\,(1+\frac{1}{t})(1-t)\,\mathrm{T}(t) = 0.$$

Thus the left side of (6.19) becomes

$$[1]\,(1+\frac{1}{t})\,(1-t)(1-\frac{Q}{t})\,\mathrm{T}(t)$$

$$(6.22) \qquad = [1]\,(1+\frac{1}{t})(1-t)\,\mathrm{T}(t) - Q\,[1]\frac{1}{t}\,(1+\frac{1}{t})(1-t)\,\mathrm{T}(t)$$

$$= -\,Q\,[1]\frac{1}{t}\,(1+\frac{1}{t})(1-t)\,\mathrm{T}(t).$$

Replacing t by $1/t$, we have

$$[1]\,(1+\frac{1}{t})\,(1-t)(1-\frac{Q}{t})\,\mathrm{T}(t) = Q\,[1]\,t\,(1+\frac{1}{t})(1-t)\,\mathrm{T}(t)$$

$$(6.23) \qquad\qquad = Q\,[1]\,(1+t)(1-t)\,\mathrm{T}(t).$$

Similarly, the constant term on the right side of (6.19) is

$$[1]\,(1+t)\,(1-t)(1-\frac{Q}{t})\,\mathrm{T}(t)$$

$$(6.24) \qquad = [1]\,(1+t)(1-t)\,\mathrm{T}(t) - Q\,[1]\,(1+\frac{1}{t})(1-t)\,\mathrm{T}(t)$$

$$= [1]\,(1+t)(1-t)\,\mathrm{T}(t).$$

The result (6.19) follows by comparing (6.23) and (6.24). $\square$

The following lemma explicitly expresses Lemma 12 in terms of $_q bc_n(a, b, k; t_1, \ldots, t_n)$.

Lemma 13. *If $\omega(t_1, \ldots, t_n)$ is invariant*

$$(6.25) \qquad \omega(t_1, \ldots, t_n) = \omega(t_1, \ldots, t_{n-1}, \frac{1}{t_n})$$

under $t_n \leftrightarrow 1/t_n$, then

$$[1]\,(1+\frac{1}{t_n})\,\omega(t_1, \ldots, t_n)\,_q bc_n(a, b, k; t_1, \ldots, t_n)$$

$$(6.26) \qquad = q^a\,[1]\,(1+t_n)\,\omega(t_1, \ldots, t_n)\,_q bc_n(a, b, k; t_1, \ldots, t_n).$$

Proof. Observe that

$$(6.27) \qquad (qt_n)_a(\frac{q}{t_n})_a = (1-t_n)(1-\frac{q^a}{t_n})\,(qt_n)_{a-1}(\frac{q}{t_n})_{a-1}.$$

By the definition (3.46) of $_q\epsilon_n(a, b, k; t_1, \ldots, t_n)$, we have

$$(6.28) \qquad \omega(t_1, \ldots, t_n)\,_q bc_n(a, b, k; t_1, \ldots, t_n) = (1-t_n)(1-\frac{q^a}{t_n})\,\mathrm{T}(t_n),$$

where

(6.29) $T(t_n) = \omega(t_1,\dots,t_n)\,(qt_n)_{a-1}(\frac{q}{t_n})_{a-1}\,{}_q\epsilon_n(a,b,k;t_1,\dots,t_n).$

By the symmetry Lemma 5 (3.48) of ${}_q\epsilon_n(a,b,k;t_1,\dots,t_n)$, we see that $T(t_n)$ is invariant under $t_n \leftrightarrow 1/t_n$. Thus we may apply Lemma 12 with $T(t_n)$ given by (6.29) and $t = t_n$, $Q = q^a$. Observe that (6.19) then becomes the result (6.26). $\square$

Lemmas 11 and 13 constitute the q-transportation theory for B_n.

The following lemma extends Lemma 10 from A_{n-1} to C_n.

Lemma 14. *Let*

(6.30) $S(s) = S(\frac{q}{s})$

be invariant under $s \leftrightarrow q/s$ *and have a Laurent expansion at* $s = 0$. *Then*

(6.31) $[1]\,s\,(1 - \frac{q}{s^2})(1 - Qs^2)\,S(s) = q^2 Q\,[1]\,\frac{1}{s}\,(1 - \frac{q}{s^2})(1 - Qs^2)\,S(s).$

Proof. Since $s - q/s$ changes sign under $s \leftrightarrow q/s$, we have

(6.32) $[1]\,(s - \frac{q}{s})\,S(s) = 0.$

Thus the left side of (6.31) becomes

$$[1]\,s\,(1 - \frac{q}{s^2})(1 - Qs^2)\,S(s)$$

(6.33)
$$= [1]\,(s - \frac{q}{s})\,S(s) - Q\,[1]\,s^2\,(s - \frac{q}{s})\,S(s)$$

$$= -Q\,[1]\,s^2\,(s - \frac{q}{s})\,S(s).$$

Replacing s by q/s, we have

(6.34) $[1]\,s\,(1 - \frac{1}{s^2})(1 - qs^2)\,S(s) = q^2 Q\,[1]\,\frac{1}{s^2}\,(s - \frac{q}{s})\,S(s).$

Similarly, the constant term on the right side of (6.31) is

$$[1]\,\frac{1}{s}\,(1 - \frac{q}{s^2})(1 - Qs^2)\,S(s)$$

(6.35)
$$= [1]\,\frac{1}{s}\,(1 - \frac{q}{s^2})\,S(s) - Q\,[1]\,(s - \frac{q}{s})\,S(s)$$

$$= [1]\,\frac{1}{s^2}\,(s - \frac{q}{s})\,S(s).$$

The result (6.31) follows by comparing (6.34) and (6.35). $\square$

The following lemma explicitly expresses Lemma 14 in terms of ${}_q bc_n(a,b,k;\ t_1,\dots,t_n)$.

Lemma 15. *If* $\omega(t_1,\dots,t_n)$ *is invariant*

$$(6.36) \qquad \omega(t_1,\dots,t_n) = \omega(\frac{q}{t_1},t_2,\dots,t_n)$$

under $t_1 \leftrightarrow q/t_1$, *then*

$$(6.37) \qquad \begin{aligned} &[1]\, t_1\, \omega(t_1,\dots,t_n)\, {}_q bc_n(a,b,k;t_1,\dots,t_n) \\ &= q^{2b+1}\, [1]\, \frac{1}{t_1}\, \omega(t_1,\dots,t_n)\, {}_q bc_n(a,b,k;t_1,\dots,t_n). \end{aligned}$$

Proof. Observe that

$$(6.38) \qquad (qt_1^2;q^2)_b (\frac{q^3}{t_1^2};q^2)_b = (1-\frac{q}{t_1^2})(1-q^{2b-1}t_1^2)\,(qt_1^2;q^2)_{b-1}(\frac{q^3}{t_1^2};q^2)_{b-1}.$$

By the definition (3.46) of $_q f_n(a,b,k;t_1,\dots,t_n)$, we have

$$(6.39) \qquad \omega(t_1,\dots,t_n)\, {}_q bc_n(a,b,k;t_1,\dots,t_n) = (1-\frac{q}{t_1^2})(1-q^{2b-1}t_1^2)\, S(t_1),$$

where

$$(6.40) \qquad S(t_1) = \omega(t_1,\dots,t_n)\,(qt_1^2;q^2)_{b-1}(\frac{q^3}{t_1^2};q^2)_{b-1}\, {}_q f_n(a,b,k;t_1,\dots,t_n).$$

By the symmetry Lemma 5 (3.49) of $_q f_n(a,b,k;t_1,\dots,t_n)$, we see that $S(t_1)$ is invariant under $t_1 \leftrightarrow q/t_n$. Thus we may apply Lemma 14 with $S(t_1)$ given by (6.40) and $s = t_1$, $Q = q^{2b-1}$. Observe that (6.31) then becomes the result (6.37). $\square$

Lemmas 11 and 15 constitute the q-transportation theory for C_n. Lemmas 11, 13 and 15 constitute the q-transportation theory for BC_n.

7. Evaluation of the constant terms A, E, K, F and Z

In this section, we complete the evaluation of the constant terms occurring in Lemma 8 (4.23) by applying the q-transportation theory for C_n, Lemmas 11 and 15, to Lemma 9 (5.2)–(5.6). We require Corollary 16, which is a technical corollary of Lemmas 11 and 15. We obtain Lemma 17 which gives q-analogues of (2.40) and (2.46)–(2.51).

The following is a technical corollary of the q-transportation theory for C_n, Lemmas 11 and 15.

Corollary 16. *Let $r \geq 1$ and let $\omega(t_1,\dots,t_n)$ be symmetric in $t_1,\dots,t_r$ and invariant under $t_1 \leftrightarrow q/t_1$. Then*

$$[1]\,\frac{1}{t_r}\,\prod_{i=1}^{r-1} t_i\,\omega(t_1,\dots,t_n)\,_q bc_n(a,b,k;t_1,\dots,t_n)$$

$$(7.1) \qquad = (q^{-k}-q^{(r-2)k})\,[1]\,\prod_{i=1}^{r-2} t_i\,\omega(t_1,\dots,t_n)\,_q bc_n(a,b,k;t_1,\dots,t_n)$$

$$+\,q^{-2b-1-(r-1)k}\,[1]\,\prod_{i=1}^{r} t_i\,\omega(t_1,\dots,t_n)\,_q bc_n(a,b,k;t_1,\dots,t_n).$$

Proof. We proceed by induction on r. For $r = 1$, we see that the first term on the right side of (7.1) is 0 and that (7.1) holds by Lemma 15 (6.37). Let $r > 1$ and assume that (7.1) holds with r replaced by $r - 1$. Replacing r by $r - 1$ and introducing an extra factor of t_r in (7.1), we have

$$[1]\,\frac{t_r}{t_{r-1}}\,\prod_{i=1}^{r-2} t_i\,\omega(t_1,\dots,t_n)\,_q bc_n(a,b,k;t_1,\dots,t_n)$$

$$(7.2) \qquad = (q^{-k}-q^{(r-3)k})\,[1]\,t_r\prod_{i=1}^{r-3} t_i\,\omega(t_1,\dots,t_n)\,_q bc_n(a,b,k;t_1,\dots,t_n)$$

$$+\,q^{-2b-1-(r-2)k}\,[1]\,\prod_{i=1}^{r} t_i\,\omega(t_1,\dots,t_n)\,_q bc_n(a,b,k;t_1,\dots,t_n).$$

Apply Lemma 11 (6.12) twice with $v = r$ and $v = r - 1$ to the first term on the right side of (7.2). This is valid for $r = 2$ since the term involved is 0. We obtain

$$[1]\,\frac{t_r}{t_{r-1}}\,\prod_{i=1}^{r-2} t_i\,\omega(t_1,\dots,t_n)\,_q bc_n(a,b,k;t_1,\dots,t_n)$$

$$(7.3) \qquad = (q^{k}-q^{(r-1)k})\,[1]\,\prod_{i=1}^{r-2} t_i\,\omega(t_1,\dots,t_n)\,_q bc_n(a,b,k;t_1,\dots,t_n)$$

$$+\,q^{-2b-1-(r-2)k}\,[1]\,\prod_{i=1}^{r} t_i\,\omega(t_1,\dots,t_n)\,_q bc_n(a,b,k;t_1,\dots,t_n).$$

Take $v = r$ and introduce an extra factor of $\prod_{i=1}^{r-2} t_i$ in Lemma 11 (6.14). This gives

$$[1] \frac{t_{r-1}}{t_r} \prod_{i=1}^{r-2} t_i \, \omega(t_1, \dots, t_n) \, {}_q bc_n(a, b, k; t_1, \dots, t_n)$$

$$(7.4) \qquad = (q^{-k} - 1)[1] \prod_{i=1}^{r-2} t_i \, \omega(t_1, \dots, t_n) \, {}_q bc_n(a, b, k; t_1, \dots, t_n)$$

$$+ q^{-k}[1] \frac{t_r}{t_{r-1}} \prod_{i=1}^{r-2} t_i \, \omega(t_1, \dots, t_n) \, {}_q bc_n(a, b, k; t_1, \dots, t_n).$$

The result (7.1) follows by substituting (7.3) into (7.4) and simplifying. $\square$

The following lemma completes the evaluation of the constant terms occurring in Lemma 8 (4.23), giving q-analogues of (2.40) and (2.46)–(2.51).

Lemma 17. *Let* $r \geq 1$. *Then*

$$(7.5) \qquad {}_q A^v_{n,0,r}(a, b, k) = 0, \quad 2 \leq v \leq r,$$

$$(7.6)$$

$$\quad {}_q A^v_{n,0,r}(a, b, k) = q^{(v-r-1)k} \, {}_q BC_{n,0,r}(a, b, k), \quad r < v \leq n,$$

$$\quad {}_q E_{n,0,r}(a, b, k) = \frac{(q^{(n-r)k} + q^{a-2b-1-(n-1)k})}{(1 + q^{2a})} \, {}_q BC_{n,0,r}(a, b, k)$$

$$(7.7) \qquad\qquad + \frac{(1 + q^a)}{(1 + q^{2a})} \, {}_q BC_{n,0,r-1}(a, b, k)$$

$$\qquad\qquad + \frac{q^{a-(n-r)k}}{(1 + q^{2a})} \left(q^{-k} - q^{(r-2)k} \right) \, {}_q BC_{n,0,r-2}(a, b, k),$$

$$\quad {}_q K^v_{n,0,r}(a, b, k) = \left(q^{(2r-2v-1)k}(1 + q^k) - q^{(2r-v-2)k} \right) \, {}_q BC_{n,0,r-2}(a, b, k)$$

$$(7.8) \qquad\qquad + q^{-2b-1-(v-1)k} \, {}_q BC_{n,0,r}(a, b, k), \quad 2 \leq v \leq r,$$

$$\quad {}_q K^v_{n,0,r}(a, b, k) = q^{-(v-r)k} \left(q^{-k} - q^{(r-2)k} \right) \, {}_q BC_{n,0,r-2}(a, b, k)$$

$$(7.9) \qquad\qquad + q^{-2b-1-(v-1)k} \, {}_q BC_{n,0,r}(a, b, k), \quad r < v \leq n,$$

$$(7.10) \qquad {}_q F_{n,0,r}(a, b, k) = q^{-2b-1} \, {}_q BC_{n,0,r}(a, b, k),$$

$$(7.11) \qquad {}_q Z_{n,0,r}(a, b, k) = q^{-1} \, {}_q BC_{n,0,r}(a, b, k).$$

Proof. The result (7.5) is Lemma 9 (5.1). It has been repeated in order to ensure the completeness of Lemma 17.

By Lemma 11 (6.12), we have

(7.12)
$$[1]t_v \prod_{i=2}^{r} t_i \, _qbc_n(a,b,k;t_2,\ldots,t_{v-1},t_1,t_v,\ldots,t_n)$$
$$= q^k [1]t_1 \prod_{i=2}^{r} t_i \, _qbc_n(a,b,k;t_2,\ldots,t_{v-1},t_1,t_v,\ldots,t_n), \quad r < v \le n.$$

Substituting (7.12) into (5.2) gives
(7.13)
$$_qA_{n,0,r}^v(a,b,k) = [1]t_1 \prod_{i=2}^{r} t_i \, _qbc_n(a,b,k;t_2,\ldots,t_{v-1},t_1,t_v,\ldots,t_n), \quad r < v \le n.$$

Replacing $(t_2,\ldots,t_{v-1},t_1,t_v,\ldots,t_n)$ by $(t_1,\ldots,t_n)$ and applying Lemma 11 (6.12) $v - r - 1$ times, we obtain

(7.14)
$$_qA_{n,0,r}^v(a,b,k) = [1]t_{v-1} \prod_{i=1}^{r-1} t_i \, _qbc_n(a,b,k;t_1,\ldots,t_n)$$
$$= q^{(v-r-1)k} [1] \prod_{i=1}^{r} t_i \, _qbc_n(a,b,k;t_1,\ldots,t_n), \quad r < v \le n,$$

which gives the result (7.6).

Replacing $(t_2,\ldots,t_n,t_1)$ by $(t_1,\ldots,t_n)$ in (5.3), we have

(7.15) $\quad _qE_{n,0,r}(a,b,k) = \dfrac{1}{(1+q^{2a})} [1](t_n+1)(1+\dfrac{q^a}{t_n}) \prod_{i=1}^{r-1} t_i \, _qbc_n(a,b,k;t_1,\ldots,t_n).$

Expanding $(t_n + 1)(1 + q^a/t_n)$ by

(7.16)
$$(t_n + 1)(1 + \frac{q^a}{t_n}) = (1 + q^a) + (t_n + \frac{q^a}{t_n})$$

and applying Lemma 11 (6.12) and (6.13) $n - r$ times each, we obtain

(7.17)
$$_qE_{n,0,r}(a,b,k)$$

$$= \frac{(1+q^a)}{(1+q^{2a})}\,[1]\prod_{i=1}^{r-1}t_i\,_qbc_n(a,b,k;t_1,\dots,t_n)$$

$$+ \frac{1}{(1+q^{2a})}\,[1]\,(t_n+\frac{q^a}{t_n})\prod_{i=1}^{r-1}t_i\,_qbc_n(a,b,k;t_1,\dots,t_n)$$

$$= \frac{(1+q^a)}{(1+q^{2a})}\,_qBC_{n,0,r-1}(a,b,k) + \frac{q^{(n-r)k}}{(1+q^{2a})}\,[1]\prod_{i=1}^{r}t_i\,_qbc_n(a,b,k;t_1,\dots,t_n)$$

$$+ \frac{q^{a-(n-r)k}}{(1+q^{2a})}\,[1]\,\frac{1}{t_r}\prod_{i=1}^{r-1}t_i\,_qbc_n(a,b,k;t_1,\dots,t_n)$$

$$= \frac{(1+q^a)}{(1+q^{2a})}\,_qBC_{n,0,r-1}(a,b,k) + \frac{q^{(n-r)k}}{(1+q^{2a})}\,_qBC_{n,0,r}(a,b,k)$$

$$+ \frac{q^{a-(n-r)k}}{(1+q^{2a})}\,[1]\,\frac{1}{t_r}\prod_{i=1}^{r-1}t_i\,_qbc_n(a,b,k;t_1,\dots,t_n).$$

Putting $\omega(t_1,\dots,t_n)=1$ in Corollary 16 (7.1) yields
(7.18)
$$[1]\,\frac{1}{t_r}\prod_{i=1}^{r-1}t_i\,_qbc_n(a,b,k;t_1,\dots,t_n)$$

$$= (q^{-k}-q^{(r-2)k})\,[1]\prod_{i=1}^{r-2}t_i\,_qbc_n(a,b,k;t_1,\dots,t_n)$$

$$+ q^{-2b-1-(r-1)k}\,[1]\prod_{i=1}^{r}t_i\,_qbc_n(a,b,k;t_1,\dots,t_n)$$

$$= (q^{-k}-q^{(r-2)k})\,_qBC_{n,0,r-2}(a,b,k) + q^{-2b-1-(r-1)k}\,_qBC_{n,0,r}(a,b,k).$$

Substituting (7.18) into (7.17) and rearranging, we obtain the result (7.7).

Replacing t_1 by $1/t_1$ in (5.4) and using the fact that

(7.19)
$$(\frac{t_v}{t_1}+1)(1+q^k\frac{t_1}{t_v}) = \frac{(t_v+t_1)}{t_1 t_v}(t_v+q^k t_1),$$

we have

(7.20)
$$_qK^v_{n,0,r}(a,b,k) = \frac{1}{(1+q^{2k})}\,[1]\,\frac{(t_v+t_1)}{t_1 t_v}(t_v+q^k t_1)\prod_{\substack{i=2\\i\neq v}}^{r}t_i$$

$$\times\,_qbc_n(a,b,k;t_2,\dots,t_v,t_1,t_{v+1},\dots,t_n),\quad 2\le v\le r.$$

By Lemma 11 (6.12), we have
(7.21)

$$[1]\, t_1\, \frac{(t_v + t_1)}{t_1 t_v} \prod_{\substack{i=2 \\ i \neq v}}^{r} t_i \; {}_q bc_n(a,b,k; t_2, \ldots, t_v, t_1, t_{v+1}, \ldots, t_n)$$

$$= q^k\, [1]\, t_v\, \frac{(t_v + t_1)}{t_1 t_v} \prod_{\substack{i=2 \\ i \neq v}}^{r} t_i \; {}_q bc_n(a,b,k; t_2, \ldots, t_v, t_1, t_{v+1}, \ldots, t_n), \quad 2 \leq v \leq r.$$

Substituting (7.21) into (7.20) and simplifying gives
(7.22)

$${}_q K_{n,0,r}^v(a,b,k)$$

$$= [1]\, \frac{(t_v + t_1)}{t_1 t_v}\, t_v \prod_{\substack{i=2 \\ i \neq v}}^{r} t_i \; {}_q bc_n(a,b,k; t_2, \ldots, t_v, t_1, t_{v+1}, \ldots, t_n)$$

$$= [1]\, (1 + \frac{t_v}{t_1}) \prod_{\substack{i=2 \\ i \neq v}}^{r} t_i \; {}_q bc_n(a,b,k; t_2, \ldots, t_v, t_1, t_{v+1}, \ldots, t_n), \quad 2 \leq v \leq r$$

Replacing $(t_2, \ldots, t_v, t_1, t_{v+1}, \ldots, t_n)$ by $(t_1, \ldots, t_n)$, we have

$${}_q K_{n,0,r}^v(a,b,k)$$

(7.23)

$$= [1]\, (1 + \frac{t_{v-1}}{t_v}) \prod_{i=1}^{v-2} t_i \prod_{i=v+1}^{r} t_i \; {}_q bc_n(a,b,k; t_1, \ldots, t_n), \quad 2 \leq v \leq r.$$

Applying Lemma 11 (6.12) $2(r - v)$ times gives

$$[1] \prod_{i=1}^{v-2} t_i \prod_{i=v+1}^{r} t_i \; {}_q bc_n(a,b,k; t_1, \ldots, t_n)$$

(7.24)

$$= q^{2(r-v)k}\, [1] \prod_{i=1}^{r-2} t_i \; {}_q bc_n(a,b,k; t_1, \ldots, t_n)$$

$$= q^{2(r-v)k}\, {}_q BC_{n,0,r-2}(a,b,k), \quad 2 \leq v \leq r.$$

Taking $r = v$ and $\omega(t_1, \ldots, t_n) = \prod_{i=v+1}^{r} t_i$ in Corollary 16 (7.1) and using (7.24), we obtain

$$[1]\frac{t_{v-1}}{t_v}\prod_{i=1}^{v-2}t_i\prod_{i=v+1}^{r}t_i\,_qbc_n(a,b,k;t_1,\dots,t_n)$$

$$=(q^{-k}-q^{(v-2)k})[1]\prod_{i=1}^{v-2}t_i\prod_{i=v+1}^{r}t_i\,_qbc_n(a,b,k;t_1,\dots,t_n)$$

(7.25)

$$+q^{-2b-1-(v-1)k}[1]\prod_{i=1}^{v}t_i\prod_{i=v+1}^{r}t_i\,_qbc_n(a,b,k;t_1,\dots,t_n)$$

$$=(q^{(2r-2v-1)k}-q^{(2r-v-2)k})\,_qBC_{n,0,r-2}(a,b,k)$$

$$+q^{-2b-1-(v-1)k}\,_qBC_{n,0,r}(a,b,k),\quad 2\le v\le r.$$

Substituting (7.24) and (7.25) into (7.22) and simplifying, we obtain the result (7.8).

Replacing t_1 by $1/t_1$ in (5.5) and using the fact that

(7.26)
$$(\frac{t_v}{t_1}+1)\frac{1}{t_v}=\frac{1}{t_1}+\frac{1}{t_v},$$

we have

(7.27)
$$_qK_{n,0,r}^{v}(a,b,k)=\frac{1}{(1+q^k)}[1](\frac{1}{t_1}+\frac{1}{t_v})\prod_{i=2}^{r}t_i$$
$$\times\,_qbc_n(a,b,k;t_2,\dots,t_v,t_1,t_{v+1},\dots,t_n),\quad r<v\le n.$$

By Lemma 11 (6.12), we have

(7.28)
$$[1]\frac{1}{t_v}\prod_{i=2}^{r}t_i\,_qbc_n(a,b,k;t_2,\dots,t_v,t_1,t_{v+1},\dots,t_n)$$
$$=q^k[1]\frac{1}{t_1}\prod_{i=2}^{r}t_i\,_qbc_n(a,b,k;t_2,\dots,t_v,t_1,t_{v+1},\dots,t_n),\quad r<v\le n.$$

Substituting (7.28) into (7.27) gives

(7.29)
$$_qK_{n,0,r}^{v}(a,b,k)$$
$$=[1]\frac{1}{t_1}\prod_{i=2}^{r}t_i\,_qbc_n(a,b,k;t_2,\dots,t_v,t_1,t_{v+1},\dots,t_n),\quad r<v\le n.$$

Replacing $(t_2,\dots,t_v,t_1,t_{v+1},\dots,t_n)$ by $(t_1,\dots,t_n)$, we have

(7.30)
$$_qK_{n,0,r}^{v}(a,b,k)=[1]\frac{1}{t_v}\prod_{i=1}^{r-1}t_i\,_qbc_n(a,b,k;t_1,\dots,t_n),\quad r<v\le n.$$

Applying Lemma 11 (6.13) $v - r$ times gives

(7.31) $_qK_{n,0,r}^v(a,b,k) = q^{-(v-r)k}\,[1]\,\dfrac{1}{t_r}\,\displaystyle\prod_{i=1}^{r-1} t_i\ _qbc_n(a,b,k;t_1,\ldots,t_n),\quad r < v \le n.$

Substituting (7.18) into (7.31), we obtain the result (7.9).
    Replacing t_1 by $1/t_1$ in (5.6) and using the fact that

(7.32) $$\left(\dfrac{q}{t_1^2}+1\right)\dfrac{t_1}{q} = \dfrac{1}{t_1}+\dfrac{t_1}{q},$$

we have

(7.33) $_qF_{n,0,r}(a,b,k) = \dfrac{1}{(1+q^{2b})}\,[1]\left(\dfrac{1}{t_1}+\dfrac{t_1}{q}\right)\displaystyle\prod_{i=2}^{r} t_i\ _qbc_n(a,b,k;t_1,\ldots,t_n).$

Taking $\omega(t_1,\ldots,t_n) = \prod_{i=2}^{r} t_i$ in Lemma 15 (6.37) gives
(7.34)
$$[1]\,t_1\displaystyle\prod_{i=2}^{r} t_i\ _qbc_n(a,b,k;t_1,\ldots,t_n) = q^{2b+1}\,[1]\,\dfrac{1}{t_1}\displaystyle\prod_{i=2}^{r} t_i\ _qbc_n(a,b,k;t_1,\ldots,t_n).$$

Substituting (7.34) into (7.33) gives

(7.35) $_qF_{n,0,r}(a,b,k) = \dfrac{(q^{-2b-1}+q^{-1})}{(1+q^{2b})}\,[1]\,t_1\displaystyle\prod_{i=2}^{r} t_i\ _qbc_n(a,b,k;t_1,\ldots,t_n)$

and the result (7.10) follows.
    Since $r \ge 1$, we see that t_1 is a factor of $\prod_{i=1}^{r} t_i$. Replacing t_1 by t_1/q in Lemma 8 (4.28) gives

(7.36) $_qZ_{n,0,r}(a,b,k) = q^{-1}\,[1]\,\displaystyle\prod_{i=1}^{r} t_i\ _qbc_n(a,b,k;t_1,t_2,\ldots,t_n)$

and the result (7.11) follows. $\square$

8. q-analogues of some functional equations

In this section, we give q-analogues of some functional equations which establish the dependence of $_qBC_{n,m}(a,b,k)$ on m and a. Substituting Lemma 17 (7.5)–(7.11) into Lemma 8 (4.23) gives the q-analogue Lemma 18 of (2.27). Using the q-transportation theory for A_{n-1}, Lemma 11, we rearrange Lemma 18, obtaining Lemmas 19, 20 and 21, which give q-analogues of (2.26), (2.24) and (2.17), respectively. Using the q-transportation theory for B_n, Lemmas 11 and 13, we obtain the q-analogue Lemma 22 of (2.20). Lemmas 21 and 22 establish the dependence of $_qBC_{n,m}(a,b,k)$ on m and a.

The following lemma is a q-analogue of (2.27).

Lemma 18.

$$
\begin{aligned}
0 = {}& \left(1 - q^{a+2b+1+(2n-r-1)k}\right) {}_qBC_{n,0,r}(a,b,k) \\
& + q^{2b+1+(n-1)k}\left(1 - q^{a}\right){}_qBC_{n,0,r-1}(a,b,k) \\
& + q^{2b+1+(r-2)k}\left(1 - q^{(r-1)k}\right){}_qBC_{n,0,r-2}(a,b,k), \quad r \geq 1.
\end{aligned}
$$

(8.1)

Proof. Substitute Lemma 17 (7.5)–(7.11) into Lemma 4 (4.23). Collecting terms, we have
(8.2)

$$
\begin{aligned}
0 = {}& \left[\left(1 - q^{k}\right)\sum_{v=r+1}^{n} q^{(v-r-1)k} + \frac{(1 - q^{a})}{(1 + q^{2a})}\left(q^{(n-r)k} + q^{a-2b-1-(n-1)k}\right)\right. \\
& \left. + \left(1 - q^{k}\right)\sum_{v=2}^{n} q^{-2b-1-(v-1)k} + \left(1 - q^{2b}\right)q^{-2b-1} + \frac{(1 - q)}{q}\right] {}_qBC_{n,0,r}(a,b,k) \\
& + \left(1 - q^{a}\right)\frac{(1 + q^{a})}{(1 + q^{2a})}\,{}_qBC_{n,0,r-1}(a,b,k) \\
& + \left[\left(1 - q^{a}\right)\frac{q^{a-(n-r)k}}{(1 + q^{2a})}\left(q^{-k} - q^{(r-2)k}\right)\right. \\
& \quad + \left(1 - q^{k}\right)\left(\sum_{v=2}^{r} q^{(2r-2v-1)k}\left(1 + q^{k}\right) - q^{(2r-v-2)k}\right) \\
& \left. + \left(1 - q^{k}\right)\sum_{v=r+1}^{n} q^{-(v-r)k}\left(q^{-k} - q^{(r-2)k}\right)\right] {}_qBC_{n,0,r-2}(a,b,k), \quad r \geq 1.
\end{aligned}
$$

We may easily sum the geometric series. The coefficient of $_qBC_{n,0,r}(a,b,k)$ equals

(8.3)

$$(1-q^{(n-r)k}) + \frac{(1-q^a)}{(1+q^{2a})}\left(q^{(n-r)k} + q^{a-2b-1-(n-1)k}\right)$$

$$+ (q^{-2b-1-(n-1)k} - q^{-2b-1}) + (q^{-2b-1} - \frac{1}{q}) + (\frac{1}{q} - 1)$$

$$= -q^{(n-r)k}\left(1 - \frac{(1-q^a)}{(1+q^{2a})}\right) + q^{-2b-1-(n-1)k}\left(q^a\frac{(1-q^a)}{(1+q^{2a})} + 1\right)$$

$$= \frac{(1+q^a)}{(1+q^{2a})}\left(-q^{a+(n-r)k} + q^{-2b-1-(n-1)k}\right).$$

The coefficient of $_qBC_{n,0,r-2}(a,b,k)$ equals

(8.4)

$$(1-q^a)\frac{q^{a-(n-r)k}}{(1+q^{2a})}\left(q^{-k} - q^{(r-2)k}\right) + (q^{-k} - q^{(2r-3)k})$$

$$- (q^{(r-2)k} - q^{(2r-3)k}) + (q^{-(n-r)k} - 1)(q^{-k} - q^{(r-2)k})$$

$$= q^{-(n-r)k}\left(q^a\frac{(1-q^a)}{(1+q^{2a})} + 1\right)\left(q^{-k} - q^{(r-2)k}\right)$$

$$= q^{-(n-r)k}\frac{(1+q^a)}{(1+q^{2a})}\left(q^{-k} - q^{(r-2)k}\right).$$

Dividing by $(1+q^a)/(1+q^{2a})$, (8.2) becomes

(8.5)

$$0 = (-q^{a+(n-r)k} + q^{-2b-1-(n-1)k})\,_qBC_{n,0,r}(a,b,k)$$

$$+ (1-q^a)\,_qBC_{n,0,r-1}(a,b,k)$$

$$+ q^{-(n-r)k}\left(q^{-k} - q^{(r-2)k}\right)\,_qBC_{n,0,r-2}(a,b,k), \quad r \geq 1.$$

The result (8.1) follows by multiplying (8.5) by $q^{2b+1+(n-1)k}$. $\square$

Expand the factor $(1 - q^a t_{n-m})$ of $_qbc_{n,m+1,r-1}(a,b,k;t_1,\ldots,t_n)$ and use the q-transportation theory for A_{n-1}, Lemma 11 (6.12), $n - m - r$ times. This gives

(8.6)

$$_qBC_{n,m+1,r-1}(a,b,k)$$

$$= {}_qBC_{n,m,r-1}(a,b,k) - q^a\,[1]\,t_{n-m}\,_qbc_{n,m,r-1}(a,b,k;t_1,\ldots,t_n)$$

$$= {}_qBC_{n,m,r-1}(a,b,k) - q^{a+(n-m-r)k}\,_qBC_{n,m,r}(a,b,k), \quad r \geq 1,$$

which is a q-analogue of (2.25).

The following lemma is a q-analogue of (2.26).

Lemma 19.

$$0 = (1 - q^{a+2b+1+(2n-m-r-1)k}) \, _qBC_{n,m,r}(a,b,k)$$

(8.7)
$$+ q^{2b+1+(n-m-1)k} \left(q^{mk} - q^a\right) \, _qBC_{n,m,r-1}(a,b,k)$$

$$+ q^{2b+1+(r-2)k} \left(1 - q^{(r-1)k}\right) \, _qBC_{n,m,r-2}(a,b,k), \quad r \geq 1.$$

Proof. We proceed by induction on m. For $m = 0$, we see that (8.7) reduces to Lemma 18 (8.1). Assume that (8.7) holds for a given m with $m \geq 0$ for all $r \geq 1$. Replacing r by $r + 1$ in (8.7) and multiplying by $q^{a+(n-m-r-1)k}$, we have

$$0 = q^{a+(n-m-r-1)k} \left(1 - q^{a+2b+1+(2n-m-r-2)k}\right) \, _qBC_{n,m,r+1}(a,b,k)$$

(8.8)
$$+ q^{a+2b+1+(2n-2m-r-2)k} \left(q^{mk} - q^a\right) \, _qBC_{n,m,r}(a,b,k)$$

$$+ q^{a+2b+1+(n-m-2)k} \left(1 - q^{rk}\right) \, _qBC_{n,m,r-1}(a,b,k).$$

Subtracting (8.8) from (8.7), we obtain

$$0 = (1 - q^{a+2b+1+(2n-m-r-2)k})$$

$$\times \left(_qBC_{n,m,r}(a,b,k) - q^{a+(n-m-r-1)k} \, _qBC_{n,m,r+1}(a,b,k)\right)$$

$$+ q^{a+2b+1+(2n-m-r-2)k} \left(1 - q^k\right) \, _qBC_{n,m,r}(a,b,k)$$

$$+ q^{2b+1+(n-m-1)k} \left(q^{mk} - q^a\right)$$

(8.9)
$$\times \left(_qBC_{n,m,r-1}(a,b,k) - q^{a+(n-m-r-1)k} \, _qBC_{n,m,r}(a,b,k)\right)$$

$$+ q^{2b+1+(r-2)k} \left(1 - q^{(r-1)k}\right)$$

$$\times \left(_qBC_{n,m,r-2}(a,b,k) - q^{a+(n-m-r+1)k} \, _qBC_{n,m,r-1}(a,b,k)\right)$$

$$- q^{a+2b+1+(n-m-2)k} \left(1 - q^k\right) \, _qBC_{n,m,r-1}(a,b,k), \quad r \geq 1.$$

Simple cancellations now yield
(8.10)

$$0 = (1 - q^{a+2b+1+(2n-m-r-2)k})$$

$$\times \left({}_qBC_{n,m,r}(a,b,k) - q^{a+(n-m-r-1)k} {}_qBC_{n,m,r+1}(a,b,k) \right)$$

$$+ q^{2b+1+(n-m-2)k} (q^{(m+1)k} - q^a)$$

$$\times \left({}_qBC_{n,m,r-1}(a,b,k) - q^{a+(n-m-r)k} {}_qBC_{n,m,r}(a,b,k) \right)$$

$$+ q^{2b+1+(r-2)k} (1 - q^{(r-1)k})$$

$$\times \left({}_qBC_{n,m,r-2}(a,b,k) - q^{a+(n-m-r+1)k} {}_qBC_{n,m,r-1}(a,b,k) \right), \quad r \geq 1.$$

Using (8.6) with each term on the right side of (8.10), this becomes

$$0 = (1 - q^{a+2b+1+(2n-m-r-2)k}) {}_qBC_{n,m+1,r}(a,b,k)$$

(8.11)
$$+ q^{2b+1+(n-m-2)k} (q^{(m+1)k} - q^a) {}_qBC_{n,m+1,r-1}(a,b,k)$$

$$+ q^{2b+1+(r-2)k} (1 - q^{(r-1)k}) {}_qBC_{n,m+1,r-2}(a,b,k), \quad r \geq 1.$$

For $r = 1$, this is valid since the last term on the right side of (8.10) is 0. Our induction is complete since (8.11) is (8.7) with m replaced by $m + 1$. □
  Solving (8.6) for ${}_qBC_{n,m,r}(a,b,k)$ gives

(8.12)
$$\begin{aligned} & {}_qBC_{n,m,r}(a,b,k) \\ &= q^{-a-(n-m-r)k} \left({}_qBC_{n,m,r-1}(a,b,k) - {}_qBC_{n,m+1,r-1}(a,b,k) \right), \quad r \geq 1, \end{aligned}$$

which is a q-analogue of (2.23).
  The following lemma is a q-analogue of (2.24).

Lemma 20.

(8.13)
$$\begin{aligned} 0 = & (1 - q^{a+2b+1+(2n-m-r-1)k}) {}_qBC_{n,m,r}(a,b,k) \\ & - (1 - q^{2a+2b+1+2(n-m)k}) {}_qBC_{n,m-1,r}(a,b,k) \\ & - q^{a+2b+1+(n-m-1)k} (1 - q^{rk}) {}_qBC_{n,m,r-1}(a,b,k), \quad m \geq 1. \end{aligned}$$

Proof. Substituting (8.12) into (8.7) and multiplying by $-q^{a+(n-m-r)k}$, we obtain
(8.14)

$$0 = (1 - q^{a+2b+1+(2n-m-r-1)k}) \left({}_qBC_{n,m+1,r-1}(a,b,k) - {}_qBC_{n,m,r-1}(a,b,k) \right)$$

$$- q^{a+2b+1+(2n-2m-r-1)k} (q^{mk} - q^a)\, {}_qBC_{n,m,r-1}(a,b,k)$$

$$- q^{a+2b+1+(n-m-2)k} (1 - q^{(r-1)k})\, {}_qBC_{n,m,r-2}(a,b,k)$$

$$= (1 - q^{a+2b+1+(2n-m-r-1)k})\, {}_qBC_{n,m+1,r-1}(a,b,k)$$

$$- (1 - q^{2a+2b+1+(2n-2m-r-1)k})\, {}_qBC_{n,m,r-1}(a,b,k)$$

$$- q^{a+2b+1+(n-m-2)k} (1 - q^{(r-1)k})\, {}_qBC_{n,m,r-2}(a,b,k), \quad r \geq 1.$$

Solve (8.6) for ${}_qBC_{n,m,r-1}(a,b,k)$ and replace r by $r-1$ in the result. This gives
(8.15)
$${}_qBC_{n,m,r-2}(a,b,k)$$
$$= {}_qBC_{n,m+1,r-2}(a,b,k) + q^{a+(n-m-r+1)k}\, {}_qBC_{n,m,r-1}(a,b,k), \quad r \geq 2.$$

Substitute (8.15) into (8.14). This is valid for $r = 1$ since the last term on the right side of (8.14) is 0. We obtain

$$0 = (1 - q^{a+2b+1+(2n-m-r-1)k})\, {}_qBC_{n,m+1,r-1}(a,b,k)$$

$$- (1 - q^{2a+2b+1+(2n-2m-r-1)k})\, {}_qBC_{n,m,r-1}(a,b,k)$$

$$- q^{a+2b+1+(n-m-2)k} (1 - q^{(r-1)k})$$

(8.16)
$$\times \left({}_qBC_{n,m+1,r-2}(a,b,k) + q^{a+(n-m-r+1)k}\, {}_qBC_{n,m,r-1}(a,b,k) \right)$$

$$= (1 - q^{a+2b+1+(2n-m-r-1)k})\, {}_qBC_{n,m+1,r-1}(a,b,k)$$

$$- (1 - q^{2a+2b+1+2(n-m-1)k})\, {}_qBC_{n,m,r-1}(a,b,k)$$

$$- q^{a+2b+1+(n-m-2)k} (1 - q^{(r-1)k})\, {}_qBC_{n,m+1,r-2}(a,b,k), \quad r \geq 1.$$

Replacing m, r, by $m-1$, $r+1$, in (8.16) gives the result (8.13). $\square$

The following lemma is a q-analogue of (2.17).

Lemma 21.

(8.17) $\quad {}_qBC_{n,m}(a,b,k) = \dfrac{(1 - q^{2a+2b+1+2(n-m)k})}{(1 - q^{a+2b+1+(2n-m-1)k})}\, {}_qBC_{n,m-1}(a,b,k), \quad m \geq 1.$

Proof. The $r = 0$ case of Lemma 18 (8.13) is

(8.18)
$$0 = (1 - q^{a+2b+1+(2n-m-1)k})\, {}_qBC_{n,m}(a,b,k)$$
$$- (1 - q^{2a+2b+1+2(n-m)k})\, {}_qBC_{n,m-1}(a,b,k), \quad m \geq 1,$$

which gives the result (8.17). $\quad\square$

Lemma 21 establishes the dependence of ${}_qBC_{n,m}(a,b,k)$ on m. Repeated applications of (8.17) gives

$$(8.19) \quad {}_qBC_{n,m}(a,b,k) = \prod_{i=1}^{m} \frac{(1-q^{2a+2b+1+2(n-i)k})}{(1-q^{a+2b+1+(2n-i-1)k})} \, {}_qBC_n(a,b,k), \quad m \ge 1,$$

which is a q-analogue of (2.18).

The following lemma provides a q-analogue of (2.20).

Lemma 22.

$$(8.20)
\begin{aligned}
[1] \prod_{i=1}^{n} (1 - \frac{A_i}{t_i})(1 - q^a t_i) \, {}_qbc_n(a,b,k;t_1,\dots,t_n) \\
= \prod_{i=1}^{n} (1 + q^{(n-i)k} A_i) \, {}_qBC_{n,n}(a,b,k).
\end{aligned}$$

Proof. Let $\omega(t_1,\dots,t_n)$ be invariant (6.25) under $t_n \leftrightarrow 1/t_n$. Subtracting

$$(8.21) \qquad\qquad (1+q^a)[1]\,\omega(t_1,\dots,t_n)\,{}_qbc_n(a,b,k;t_1,\dots,t_n)$$

from both sides of Lemma 13 (6.26), we obtain

$$(8.22)
\begin{aligned}
[1]\frac{1}{t_n}(1-q^a t_n)\,\omega(t_1,\dots,t_n)\,{}_qbc_n(a,b,k;t_1,\dots,t_n) \\
= -[1](1-q^a t_n)\,\omega(t_1,\dots,t_n)\,{}_qbc_n(a,b,k;t_1,\dots,t_n).
\end{aligned}$$

Multiplying (8.22) by $-A$ and adding

$$(8.23) \qquad\qquad [1](1-q^a t_n)\,\omega(t_1,\dots,t_n)\,{}_qbc_n(a,b,k;t_1,\dots,t_n),$$

we have

$$(8.24)
\begin{aligned}
[1](1-\frac{A}{t_n})(1-q^a t_n)\,\omega(t_1,\dots,t_n)\,{}_qbc_n(a,b,k;t_1,\dots,t_n) \\
= (1+A)[1](1-q^a t_n)\,\omega(t_1,\dots,t_n)\,{}_qbc_n(a,b,k;t_1,\dots,t_n).
\end{aligned}$$

Apply the q-transportation theory for A_{n-1}, Lemma 11 (6.13), $n-m$ times with v running from $m+1$ to n to the contribution of the term A_m/t_m of the factor $(1 - A_m/t_m)$ and then use (8.24). We obtain

$$(8.25)$$

$$\begin{aligned}
[1] \prod_{i=1}^{m} (1 - \frac{A_i}{t_i}) \prod_{i=1}^{n} (1 - q^a t_i) \, {}_qbc_n(a,b,k;t_1,\dots,t_n) \\
= [1] \prod_{i=1}^{m-1} (1 - \frac{A_i}{t_i})(1 - q^{(n-m)k}\frac{A_m}{t_n}) \prod_{i=1}^{n} (1 - q^a t_i) \, {}_qbc_n(a,b,k;t_1,\dots,t_n) \\
= (1 + q^{(n-m)k} A_m)[1] \prod_{i=1}^{m-1} (1 - \frac{A_i}{t_i}) \prod_{i=1}^{n} (1 - q^a t_i) \, {}_qbc_n(a,b,k;t_1,\dots,t_n).
\end{aligned}$$

Repeated applications of (8.25) gives the result (8.20). $\square$

Lemmas 21 and 22 establish the dependence of $_qBC_{n,m}(a,b,k)$ on m and a. Taking $A_i = q^{a+1}$ in (8.20) gives

$$\begin{aligned}
(8.26) \quad _qBC_n(a+1,b,k) &= [1] \prod_{i=1}^{n}(1 - \frac{q^{a+1}}{t_i})(1 - q^a t_i) \, _qbc_n(a,b,k;t_1,\ldots,t_n) \\
&= \prod_{i=1}^{n}(1 + q^{a+1+(n-i)k}) \, _qBC_{n,n}(a,b,k),
\end{aligned}$$

which is a q-analogue of (2.20).

The $m = n$ case of (8.19) is

$$(8.27) \qquad _qBC_{n,n}(a,b,k) = \prod_{i=1}^{n} \frac{(1 - q^{2a+2b+1+2(n-i)k})}{(1 - q^{a+2b+1+(2n-i-1)k})} \, _qBC_n(a,b,k).$$

Substituting (8.27) into (8.26), we have
(8.28)
$$_qBC_n(a+1,b,k) = \prod_{i=1}^{n}(1 + q^{a+1+(n-i)k)}) \frac{(1 - q^{2a+2b+1+2(n-i)k)})}{(1 - q^{a+2b+1+(2n-i-1)k})} \, _qBC_n(a,b,k),$$

which is a q-analogue of (2.21).

9. q-transportation theory revisited

In this section, we show how to obtain the q-analogues Lemmas 18 (8.1) and 21 (8.17) of (2.27) and (2.17), respectively, using only the q-transportation theory for BC_n, Lemmas 11, 13 and 15. This is based upon the fact that B_n and C_n have the same Weyl group. Since Lemmas 21 and 22 are the only results which we require to complete the proof of Theorem 4 (1.18), we may supplant the q-engine (4.4) of our q-machine by the q-transportation theory for BC_n.

The Weyl groups of the root systems B_n and C_n are both isomorphic to $2^n \ltimes S_n$, which consists of the signed permutations $\langle \pi, \epsilon \rangle$, where $\pi \in S_n$ and $\epsilon(i) = \pm 1$, $1 \leq i \leq n$. Observe that the q-transportation theory for A_{n-1}, Lemma 11 (6.12)–(6.14), may be extended to a q-analogue of the transportation property (2.14) for BC_n by either Lemma 13 (6.26) or Lemma 15 (6.37). We may equate the results of these two approaches. Observe that each of the parameters k, a and b, which are associated with the roots of A_{n-1}, B_n and C_n, with length $\sqrt{2}$, 1 and 2, respectively, occur in and only in Lemmas 11, 13 and 15, respectively. Thus we may obtain some nontrivial functional equations using only the q-transportation theory for BC_n. We now prove Lemmas 18 (8.1) and 21 (8.17) in this way.

Let us apply our idea in place of the q-engine (4.4) of our q-machine to $_qbc_{n,0,r}(a, b, k; t_1, \ldots, t_n)$, $r \geq 1$. Using the q-transportation theory for A_{n-1}, Lemma 11 (6.12), $n - r$ times, we have

(9.1)
$$_qBC_{n,0,r}(a, b, k) = [1] \prod_{i=1}^{r} t_i \, _qbc_n(a, b, k; t_1, \ldots, t_n)$$
$$= q^{-(n-r)k} [1] t_n \prod_{i=1}^{r-1} t_i \, _qbc_n(a, b, k; t_1, \ldots, t_n).$$

Using the q-transportation theory for B_n, Lemma 13 (6.26), we obtain

(9.2)
$$_qBC_{n,0,r}(a, b, k) = q^{-(n-r)k} [1] \left((1 + t_n) - 1 \right) \prod_{i=1}^{r-1} t_i \, _qbc_n(a, b, k; t_1, \ldots, t_n)$$
$$= q^{-(n-r)k} [1] \left(q^{-a}(1 + \frac{1}{t_n}) - 1 \right) \prod_{i=1}^{r-1} t_i \, _qbc_n(a, b, k; t_1, \ldots, t_n).$$

Simplifying and using the q-transportation theory for A_{n-1}, Lemma 11 (6.13), $n - r$ times yields

(9.3)
$$_qBC_{n,0,r}(a, b, k) = q^{-(n-r)k} (q^{-a} - 1) [1] \prod_{i=1}^{r-1} t_i \, _qbc_n(a, b, k; t_1, \ldots, t_n)$$
$$+ q^{-(n-r)k-a} [1] \frac{1}{t_n} \prod_{i=1}^{r-1} t_i \, _qbc_n(a, b, k; t_1, \ldots, t_n)$$
$$= q^{-(n-r)k} (q^{-a} - 1) \, _qBC_{n,0,r-1}(a, b, k)$$
$$+ q^{-2(n-r)k-a} [1] \frac{1}{t_r} \prod_{i=1}^{r-1} t_i \, _qbc_n(a, b, k; t_1, \ldots, t_n).$$

We may now change the factor $1/t_r$ to $1/t_1$ by using the q-transportation theory for A_{n-1}, Lemma 11 (6.14), $r-1$ times with v running from r down to 2 and then change the factor $1/t_1$ to t_1 by using the q-transportation theory for C_n, Lemma 15 (6.37). This may be done by using Corollary 16 (7.1). Taking $\omega(t_1, \ldots, t_n) = 1$ in (7.1) gives

$$
\begin{aligned}
[1] & \frac{1}{t_r} \prod_{i=1}^{r-1} t_i \; {}_qbc_n(a,b,k;t_1,\ldots,t_n) \\
& = (q^{-k} - q^{(r-2)k})\,[1] \prod_{i=1}^{r-2} t_i \; {}_qbc_n(a,b,k;t_1,\ldots,t_n) \\
& \quad + q^{-2b-1-(r-1)k}\,[1] \prod_{i=1}^{r} t_i \; {}_qbc_n(a,b,k;t_1,\ldots,t_n) \\
& = (q^{-k} - q^{(r-2)k})\,{}_qBC_{n,0,r-2}(a,b,k) + q^{-2b-1-(r-1)k}\,{}_qBC_{n,0,r}(a,b,k).
\end{aligned}
$$

(9.4)

Substituting (9.4) into (9.3) gives

$$
\begin{aligned}
{}_qBC_{n,0,r}(a,b,k) = &\; q^{-(n-r)k}\,(q^{-a} - 1)\,{}_qBC_{n,0,r-1}(a,b,k) \\
& + q^{-2(n-r)k-a}\,(q^{-k} - q^{(r-2)k})\,{}_qBC_{n,0,r-2}(a,b,k) \\
& + q^{-(2n-r-1)k-a-2b-1}\,{}_qBC_{n,0,r}(a,b,k).
\end{aligned}
$$

(9.5)

Multiplying (9.5) by $q^{a+2b+1+(2n-r-1)k}$ and moving the term on the left side to the right side gives Lemma 18 (8.1).

Since Lemmas 21 and 22 are the only results which we require to complete the proof of Theorem 4, we may supplant the q-engine (4.4) of our q-machine by the q-transportation theory for BC_n.

Since the partial q-derivative is linear, we expect that Lemma 21 (8.17) follows by applying the q-engine (4.4) of our q-machine to ${}_qbc_{n,m}(a,b,k;t_1,\ldots,t_n)$, $m \geq 1$. This is the case but the calculations are as tedious as our rearrangements of Lemma 18 (8.1).

Let us apply our idea in place of the q-engine (4.4) of our q-machine to ${}_qbc_{n,m}(a,b,k;t_1,\ldots,t_n)$, $m \geq 1$. Taking $\omega(t_1,\ldots,t_n) = 1$ in the formulation (8.22) of the q-transportation theory for B_n, Lemma 13 (6.26), we have

$$
\begin{aligned}
{}_qBC_{n,m}(a,b,k) = &\; [1] \prod_{i=n-m+1}^{n} (1 - q^a t_i)\,{}_qbc_n(a,b,k;t_1,\ldots,t_n) \\
& = -[1]\frac{1}{t_n} \prod_{i=n-m+1}^{n} (1 - q^a t_i)\,{}_qbc_n(a,b,k;t_1,\ldots,t_n).
\end{aligned}
$$

(9.6)

Using the q-transportation theory for A_{n-1}, Lemma 11 (6.13), $m-1$ times with v running from n to $n-m+2$ yields
(9.7)
$$_qBC_{n,m}(a,b,k)$$
$$= -q^{-(m-1)k}\,[1]\,\frac{1}{t_{n-m+1}}\prod_{i=n-m+1}^{n}(1-q^a t_i)\,_qbc_n(a,b,k;t_1,\ldots,t_n).$$

We have
(9.8)
$$\frac{1}{t_{n-m+1}}(1-q^a t_{n-m+1}) = -q^a + \frac{1}{t_{n-m+1}}.$$

Substituting (9.8) into (9.7) and using q-transportation theory for A_{n-1}, Lemma 11 (6.13), $n-m$ times with v running from $n-m+1$ to 2, we obtain

$$_qBC_{n,m}(a,b,k)$$

(9.9)
$$= q^{a-(m-1)k}\,[1]\prod_{i=n-m+2}^{n}(1-q^a t_i)\,_qbc_n(a,b,k;t_1,\ldots,t_n)$$

$$\quad - q^{-(m-1)k}\,[1]\,\frac{1}{t_{n-m+1}}\prod_{i=n-m+2}^{n}(1-q^a t_i)\,_qbc_n(a,b,k;t_1,\ldots,t_n)$$

$$= q^{a-(m-1)k}\,_qBC_{n,m-1}(a,b,k)$$

$$\quad - q^{-(n-1)k}\,[1]\,\frac{1}{t_1}\prod_{i=n-m+2}^{n}(1-q^a t_i)\,_qbc_n(a,b,k;t_1,\ldots,t_n).$$

Observe that $\prod_{i=n-m+2}^{n}(1-q^a t_i)$ is independent of t_1 since $m \le n$. We may change the factor $1/t_1$ to t_1 by using the q-transportation theory for C_n, Lemma 15 (6.37). We obtain
(9.10)
$$_qBC_{n,m}(a,b,k) = q^{a-(m-1)k}\,_qBC_{n,m-1}(a,b,k)$$

$$\quad - q^{-2b-1-(n-1)k}\,[1]\,t_1\prod_{i=n-m+2}^{n}(1-q^a t_i)\,_qbc_n(a,b,k;t_1,\ldots,t_n).$$

We may change the factor t_1 to t_{n-m+1} by using the q-transportation theory for A_{n-1}, Lemma 11 (6.12), $n-m$ times with v running from 2 to $n-m+1$. This gives
(9.11)
$$_qBC_{n,m}(a,b,k)$$

$$= q^{a-(m-1)k}\,_qBC_{n,m-1}(a,b,k)$$

$$\quad - q^{-2b-1-(2n-m-1)k}\,[1]\,t_{n-m+1}\prod_{i=n-m+2}^{n}(1-q^a t_i)\,_qbc_n(a,b,k;t_1,\ldots,t_n).$$

We have

$$(9.12) \qquad t_{n-m+1} = -q^{-a}\left(1 - q^a t_{n-m+1}\right) + q^{-a}.$$

Substituting (9.12) into (9.11), we obtain
(9.13)
$$_qBC_{n,m}(a,b,k)$$

$$= q^{a-(m-1)k}\,_qBC_{n,m-1}(a,b,k)$$

$$+ q^{-a-2b-1-(2n-m-1)k}\,[1] \prod_{i=n-m+1}^{n}(1-q^a t_i)\,_qbc_n(a,b,k;t_1,\dots,t_n)$$

$$- q^{-a-2b-1-(2n-m-1)k}\,[1] \prod_{i=n-m+2}^{n}(1-q^a t_i)\,_qbc_n(a,b,k;t_1,\dots,t_n)$$

$$= q^{a-(m-1)k}\,_qBC_{n,m-1}(a,b,k) + q^{-a-2b-1-(2n-m-1)k}\,_qBC_{n,m}(a,b,k)$$

$$- q^{-a-2b-1-(2n-m-1)k}\,_qBC_{n,m-1}(a,b,k).$$

Multiplying (9.13) by $q^{a+2b+1+(2n-m-1)k}$ and rearranging gives Lemma 21 (8.17).

10. A proof of Theorem 4

In this section, we complete the proof of Theorem 4 (1.18). We analytically continue $_qBC_{n,n}(a,b,k)$ as a function of a and, using the $a = b = 0$ case of Theorem 3 (1.14), we obtain a q-analogue of (2.22). Theorem 4 follows by analytically continuing the product on the right side of (1.18) as a function of a and comparing the result for $m = n$, $a = -1 - 2b - (n-1)k$.

We have the well-known (see Andrews [An1, (2.2.1)]) q-binomial theorem

$$(10.1) \qquad \frac{(at)_\infty}{(t)_\infty} = \sum_{i=0}^{\infty} \frac{(a)_i}{(q)_i} t^i,$$

which holds provided that $|t| < 1$ or that the sum on the right side of (10.1) terminates. Let M and N be nonnegative integers. The $a = q^{-M}$, $t = q^M t$, case of (10.1) is usually given by

$$(10.2) \qquad (t)_M = \sum_{i=0}^{M} (-1)^i q^{\binom{i}{2}} \begin{bmatrix} M \\ i \end{bmatrix} t^i,$$

where

$$(10.3) \qquad \begin{bmatrix} M \\ i \end{bmatrix} = \begin{bmatrix} M \\ i \end{bmatrix}_q = \frac{(q^{M+1-i})_i}{(q)_i}, \qquad 0 \le i \le M,$$

denotes the q-binomial coefficient. For M a nonnegative integer and i an integer, we usually set $\begin{bmatrix} M \\ i \end{bmatrix} = 0$ when $i < 0$ or $M < i$. We see that (10.3) gives the Laurent expansion of a polynomial with only simple zeros which form a geometric sequence. Observe that t^N times

$$(10.4) \qquad (t)_M \left(\frac{q}{t}\right)_N = \sum_{i=-N}^{M} (-1)^i q^{\binom{i}{2}} \begin{bmatrix} M+N \\ N+i \end{bmatrix} t^i$$

is such a polynomial, so that (10.4) is equivalent to (10.2). See Kadell [Ka1]. Since $\binom{-i}{2} = \binom{i+1}{2}$, we have

$$(10.5) \qquad [1] t^z \, (t)_M \left(\frac{q}{t}\right)_N = (-1)^z q^{\binom{z+1}{2}} \begin{bmatrix} M+N \\ N-z \end{bmatrix},$$

where z is an integer.

Fix nonnegative integers b and k. We have the polynomial
(10.6)

$$_qbc_n(0,b,k;t_1,\ldots,t_n) = \prod_{i=1}^{n} (qt_i^2;q^2)_b \left(\frac{q}{t_i^2};q^2\right)_b \prod_{1 \le i < j \le n} \left(\frac{t_i}{t_j}\right)_k \left(q\frac{t_j}{t_i}\right)_k (t_i t_j)_k \left(\frac{q}{t_i t_j}\right)_k$$

$$= \sum_{\alpha} {_q\mathcal{D}_n(b,k;\alpha)} \prod_{i=1}^{n} t_i^{\alpha_i}.$$

We may use (10.5) and (10.6) to extract the constant term $_qBC_{n,m}(a,b,k)$ in the Laurent expansion of $_qbc_{n,m}(a,b,k;t_1,\dots,t_n)$. We obtain
(10.7)

$$_qBC_{n,m}(a,b,k) = [1] \prod_{i=1}^{n}(t_i)_{a+\chi(n-i+1\leq m)}(\frac{q}{t_i})_a \; _qbc_n(0,b,k;t_1,\dots,t_n)$$

$$= \sum_{\alpha} {}_q\mathcal{D}_n(b,k;\alpha) \prod_{i=1}^{n}[1]\,t_i^{\alpha_i}\,(t_i)_{a+\chi(n-i+1\leq m)}(\frac{q}{t_i})_a$$

$$= \sum_{\alpha} {}_q\mathcal{D}_n(b,k;\alpha) \prod_{i=1}^{n}(-1)^{\alpha_i}\,q^{\binom{\alpha_i+1}{2}} \begin{bmatrix} 2a+\chi(n-i+1\leq m) \\ a-\alpha_i \end{bmatrix}.$$

We require only the $m=n$ case. It is

(10.8) $$\qquad {}_qBC_{n,n}(a,b,k) = \sum_{\alpha} {}_q\mathcal{D}_n(b,k;\alpha) \prod_{i=1}^{n}(-1)^{\alpha_i}\,q^{\binom{\alpha_i+1}{2}} \begin{bmatrix} 2a+1 \\ a-\alpha_i \end{bmatrix}.$$

Observe that since $_qbc_n(0,b,k;t_1,\dots,t_n)$ is a polynomial, (10.7) and (10.8) are finite sums.

For fixed real q with $0 < q < 1$, we let $\ell n\, q$ denote the real natural logarithm of q. We see that

(10.9) $$q^a = \exp(a\,\ell n\, q)$$

is an entire function of a. Observe that

(10.10) $$(x)_n = \frac{(x)_\infty}{(xq^n)_\infty}$$

extends the definition (1.11) of $(x)_n$ to all complex n. Using (10.10), we see that (10.3) becomes

(10.11) $$\begin{bmatrix} 2a+1 \\ a-z \end{bmatrix} = \frac{(q^{a+2+z})_{a-z}}{(q)_{a-z}} = \frac{(q^{a+1-z})_\infty}{(q)_\infty}\,\frac{(q^{a+2+z})_\infty}{(q^{2a+2})_\infty}.$$

Let z be a fixed integer. Since the partial products converge uniformly in any compact region, $(q^{a+z})_\infty$ is an entire function of q^a and hence is an entire function of a. Similarly, $(q^{2a+2})_\infty$ is an entire function of a. Thus, (10.11) gives the analytic continuation of $\begin{bmatrix} 2a+1 \\ a-z \end{bmatrix}$ as a function of a. Observe that $(q^{2a+2})_\infty$ has simple zeros when a is a negative integer. In this case we have $a+1 \leq 0$ and hence either $z \geq 0$ and $-a+z-1 \geq 0$ or $z+1 \leq 0$ and $-a-2-z \geq 0$. Thus, either $(q^{a+1-z})_\infty$ or $(q^{a+2+z})_\infty$ has a simple zero. Thus $\begin{bmatrix} 2a+1 \\ a-z \end{bmatrix}$ is an entire function of a.

Setting $z = \alpha_i$, we see that (10.11) provides the analytic continuation of the q-binomial coefficient $\begin{bmatrix} 2a+1 \\ a-\alpha_i \end{bmatrix}$ appearing in (10.8). Since (10.8) is a finite sum, this gives the analytic continuation of $_qBC_{n,n}(a,b,k)$ as a function of a. Observe that $_qBC_{n,n}(a,b,k)$ is an entire function of a.

Let $N \geq 0$. We investigate the analytic continuation of $\begin{bmatrix} 2a+1 \\ a-z \end{bmatrix}$ at the negative integer $a = -1 - N$. The product $(q^{2a+2})_\infty$ in the denominator of (10.11) has a simple zero. If $-N \leq z \leq N - 1$, then $-N - z \leq 0$ and $1 - N + z \leq 0$. Thus, both of the products $(q^{a+1-z})_\infty$ and $(q^{a+2+z})_\infty$ in the numerator of (10.11) have simple zeros. We have

$$(10.12) \qquad \begin{bmatrix} 2a+1 \\ a-z \end{bmatrix} \Bigg|_{a=-1-N} = 0 \text{ if } -N \leq z \leq N - 1.$$

We require only the $z = N$ case. It is

$$(10.13)$$
$$\begin{bmatrix} 2a+1 \\ a-N \end{bmatrix} \Bigg|_{a=-1-N} = \lim_{\epsilon \to 0} \frac{(q^{\epsilon+1})_\infty \ (q^{\epsilon-2N})_\infty}{(q)_\infty \ (q^{2\epsilon-2N})_\infty} = \lim_{\epsilon \to 0} \frac{(1-q^\epsilon)}{(1-q^{2\epsilon})} = \frac{1}{2}, \quad N \geq 0.$$

Observe that

$$(10.14)$$
$$(qt_i^2; q^2)_b \left(\frac{q}{t_i^2}; q^2\right)_b$$

$$= (1 - qt_i^2)(1 - q^3 t_i^2) \cdots (1 - q^{2b-1} t_i^2)\left(1 - \frac{q}{t_i^2}\right)\left(1 - \frac{q^3}{t_i^2}\right) \cdots \left(1 - \frac{q^{2b-1}}{t_i^2}\right)$$

$$= (-qt_i^2)(-q^3 t_i^2) \cdots (-q^{2b-1} t_i^2) + \ldots + \left(-\frac{q}{t_i^2}\right)\left(-\frac{q^3}{t_i^2}\right) \cdots \left(-\frac{q^{2b-1}}{t_i^2}\right)$$

$$= (-1)^b q^{b^2} t_i^{2b} + \ldots + (-1)^b q^{b^2} t_i^{-2b}.$$

We can give the complete expansion by replacing q by q^2 and t by qt_i^2 in the formulation (10.4) of the q-binomial theorem. Similarly, we have

$$(10.15) \qquad \left(\frac{t_i}{t_j}\right)_k \left(q\frac{t_j}{t_i}\right)_k = (-1)^k q^{\binom{k}{2}} \left(\frac{t_i}{t_j}\right)^k + \ldots + (-1)^k q^{\binom{k+1}{2}} \left(\frac{t_j}{t_i}\right)^k$$

and

$$(10.16) \qquad (t_i t_j)_k \left(\frac{q}{t_i t_j}\right)_k = (-1)^k q^{\binom{k}{2}} (t_i t_j)^k + \ldots + (-1)^k q^{\binom{k+1}{2}} (t_i t_j)^{-k}.$$

Let us now evaluate the analytic continuation of ${}_q BC_{n,n}(a, b, k)$ as a function of a at $a = -1 - 2b - (n-1)k$. Let α be chosen so that the coefficient ${}_q \mathcal{D}_n(b, k; \alpha)$ in the expansion (10.6) of ${}_q bc_n(0, b, k; t_1, \ldots, t_n)$ is not 0

$$(10.17) \qquad\qquad {}_q \mathcal{D}_n(b, k; \alpha) \neq 0$$

and so that the product of the q-binomial coefficients in the sum (10.8) for ${}_q BC_{n,n}(a, b, k)$ evaluated at $a = -1 - 2b - (n-1)k$ is not 0

$$(10.18) \qquad \prod_{i=1}^{n} \begin{bmatrix} 2a+1 \\ a-\alpha_i \end{bmatrix} \Bigg|_{a=-1-2b-(n-1)k} \neq 0.$$

For $1 \leq i \leq n$, the maximum contribution of (10.14) to $|\alpha_i|$ is $2b$. Let $1 \leq i < j \leq n$. If α_i and α_j have the same sign, then (10.15) can make no contribution to $|\alpha_i| + |\alpha_j|$, while (10.16) can contribute at most $2k$. If α_i and α_j have opposite signs, then (10.16) can make no contribution to $|\alpha_i| + |\alpha_j|$, while (10.15) can contribute at most $2k$. Thus (10.17) requires that

(10.19)
$$\sum_{i=1}^{n} |\alpha_i| \leq \sum_{i=1}^{n} 2b + \sum_{1 \leq i < j \leq n} 2k$$
$$= 2nb + \binom{n}{2} 2k = n(2b + (n-1)k).$$

Setting $z = \alpha_i$, $N = 2b + (n-1)k$, in (10.12), we see that (10.18) requires that

(10.20) $\alpha_i \leq -1 - 2b - (n-1)k$ or $\alpha_i \geq 2b + (n-1)k$ for all i, $1 \leq i \leq n$.

Comparing (10.19) and (10.20), we see that

(10.21) $\alpha_i = 2b + (n-1)k$ for all i, $1 \leq i \leq n$.

Thus the sum (10.8) becomes a single sum. Using the evaluation (10.13), we obtain (10.22)

$_q BC_{n,n}(-1 - 2b - (n-1)k, b, k)$

$$= {}_q \mathcal{D}_n(b, k; 2b + (n-1)k, \ldots, 2b + (n-1)k)$$

$$\times \prod_{i=1}^{n} (-1)^{2b+(n-1)k} q^{\binom{2b+(n-1)k+1}{2}} \left[\begin{array}{c} 2a+1 \\ a - 2b - (n-1)k \end{array} \right] \Bigg|_{a=-1-2b-(n-1)k}$$

$$= {}_q \mathcal{D}_n(b, k; 2b + (n-1)k, \ldots, 2b + (n-1)k) \, q^{n\binom{2b+(n-1)k+1}{2}} \, 2^{-n}.$$

For each i, $1 \leq i \leq n$, the inequality (10.19) is just barely satisfied. This can be achieved only by taking the product over $1 \leq i \leq n$ of the first term on the right side of (10.14) times the product over $1 \leq i < j \leq n$ of the first term on the right side of (10.16). This gives

(10.23)
$$\prod_{i=1}^{n} (-1)^b \, q^{b^2} \, t_i^{2b} \prod_{1 \leq i < j \leq n} (-1)^k \, q^{\binom{k}{2}} \, (t_i t_j)^k$$
$$= (-1)^{\left(nb + \binom{n}{2}k\right)} \, q^{\left(nb^2 + \binom{n}{2}\binom{k}{2}\right)} \prod_{i=1}^{n} t_i^{2b+(n-1)k}.$$

Since each t_i, $1 \leq i \leq n$, occurs with the desired exponent $2b + (n-1)k$, we must take the constant term in the product of the remaining factors $\prod_{1 \leq i < j \leq n} (t_i/t_j)_k$ $(qt_j/t_i)_k$ of $_qbc_n(0, b, k; t_1, \ldots, t_n)$. We obtain

(10.24)
$$_q\mathcal{D}_n(b, k; 2b + (n-1)k, \ldots, 2b + (n-1)k)$$
$$= (-1)^{(nb + \binom{n}{2}k)} q^{(nb^2 + \binom{n}{2}\binom{k}{2})} [1] \prod_{1 \leq i < j \leq n} (\frac{t_i}{t_j})_k (q\frac{t_j}{t_i})_k.$$

The $a = b = 0$ case of the q-Morris theorem, Theorem 3 (1.14), is

(10.25) $$[1] \prod_{1 \leq i < j \leq n} (\frac{t_i}{t_j})_k (q\frac{t_j}{t_i})_k = \prod_{i=1}^{n} \frac{(q)_{ik}}{(q)_{(n-i)k}(q)_k} = \frac{(q)_{nk}}{(q)_k^n}.$$

Substituting (10.25) into (10.24) and the result into (10.22) gives

(10.26)
$$_qBC_{n,n}(-1 - 2b - (n-1)k, b, k)$$
$$= (-1)^{(nb + \binom{n}{2}k)} q^{(nb^2 + \binom{n}{2}\binom{k}{2} + n(2b + (n-1)k+1))} 2^{-n} \frac{(q)_{nk}}{(q)_k^n},$$

which is a q-analogue of (2.22).

Let $_q\text{Pr}_{n,m}(a, b, k)$ denote the product
(10.27)
$$_q\text{Pr}_{n,m}(a, b, k) = \prod_{i=1}^{n} \frac{(q)_{2a+2b+2(n-i)k+\chi(i \leq m)}}{(q)_{a+2b+(2n-i-1)k+\chi(i \leq m)}(q)_{a+(n-i)k}} \frac{(q)_{2b+2(n-i)k}(q)_{ik}}{(q)_{2b+(n-i)k}(q)_k}$$
$$\times \prod_{i=1}^{n} \frac{(q^2; q^2)_{2b+(n-i)k}(q^2; q^2)_{a+(n-i)k}}{(q^2; q^2)_{b+(n-i)k}(q^2; q^2)_{a+b+(n-i)k}}$$

on the right side of (1.18). We have

(10.28) $$_q\text{Pr}_{n,m}(a, b, k) = \frac{(1 - q^{2a+2b+1+2(n-m)k})}{(1 - q^{a+2b+1+(2n-m-1)k})} \, _q\text{Pr}_{n,m-1}(a, b, k), \quad m \geq 1,$$

and

(10.29)
$$_q\text{Pr}_n(a + 1, b, k) = \prod_{i=1}^{n} \frac{(1 - q^{2a+2+2(n-i)k})}{(1 - q^{a+1+(n-i)k})} \, _q\text{Pr}_{n,n}(a, b, k)$$
$$= \prod_{i=1}^{n} (1 + q^{a+1+(n-i)k}) \, _q\text{Pr}_{n,n}(a, b, k).$$

Thus $_q\mathrm{Pr}_{n,m}(a,b,k)$ satisfies (8.17) and (8.26). We see that $_qBC_{n,m}(a,b,k)$ and $_q\mathrm{Pr}_{n,m}(a,b,k)$ share the same behavior as functions of m and a. They both satisfy

$_qBC_{n,n}(a+1,b,k)$

(10.30)
$$= \prod_{i=1}^{n} \frac{(1-q^{2a+2b+3+2(n-i)k})}{(1-q^{a+2b+2+(2n-i-1)k})} \,_qBC_n(a+1,b,k)$$

$$= \prod_{i=1}^{n}(1+q^{a+1+(n-i)k}) \frac{(1-q^{2a+2b+3+2(n-i)k})}{(1-q^{a+2b+2+(2n-i-1)k})} \,_qBC_{n,n}(a,b,k),$$

which is obtained by replacing a by $a+1$ in (8.27) and substituting (8.26) into the result. In order to prove Theorem 4, it suffices to evaluate the analytic continuation of $_q\mathrm{Pr}_{n,n}(a,b,k)$ as a function of a at $a=-1-2b-(n-1)k$ and check that the result agrees with (10.26).

From (1.18), we have

(10.31)
$$_q\mathrm{Pr}_{n,n}(a,b,k) = \prod_{i=1}^{n} \frac{(q)_{2a+2b+1+2(n-i)k}}{(q)_{a+2b+1+(2n-i-1)k} \,(q)_{a+(n-i)k}} \frac{(q)_{2b+2(n-i)k}\,(q)_{ik}}{(q)_{2b+(n-i)k}\,(q)_k}$$

$$\times \prod_{i=1}^{n} \frac{(q^2;q^2)_{2b+(n-i)k}\,(q^2;q^2)_{a+(n-i)k}}{(q^2;q^2)_{b+(n-i)k}\,(q^2;q^2)_{a+b+(n-i)k}}.$$

By (10.10), we have

(10.32)
$$\frac{(q)_{2a+2b+1+2(n-i)k}}{(q)_{a+2b+1+(2n-i-1)k}\,(q)_{a+(n-i)k}}$$

$$= \frac{(q^{a+2b+2+(2n-i-1)k})_{\infty}\,(q^{a+1+(n-i)k})_{\infty}}{(q)_{\infty}} \frac{(q)_{\infty}}{(q^{2a+2b+2+2(n-i)k})_{\infty}}$$

$$= \frac{(q^{a+2b+2+(2n-i-1)k})_{\infty}\,(q^{a+1+(n-i)k})_{\infty}}{(q)_{\infty}} \frac{1}{(q^{2a+2b+2+2(n-i)k})_{\infty}}$$

and

(10.33)
$$(x)_{n+m} = \frac{(x)_{\infty}}{(xq^{n+m})_{\infty}} = \frac{(x)_{\infty}}{(xq^n)_{\infty}} \frac{(xq^n)_{\infty}}{(xq^{n+m})_{\infty}} = (x)_n(xq^n)_m.$$

This gives

(10.34)
$$\frac{(q^2;q^2)_{2b+(n-i)k}\,(q^2;q^2)_{a+(n-i)k}}{(q^2;q^2)_{b+(n-i)k}\,(q^2;q^2)_{a+b+(n-i)k}} = \frac{(q^{2b+2+2(n-i)k};q^2)_b}{(q^{2a+2+2(n-i)k};q^2)_b}.$$

Thus (10.31) becomes

(10.35)
$_q\mathrm{Pr}_{n,n}(a,b,k)$

$$= \prod_{i=1}^{n} \frac{(q^{a+2b+2+(2n-i-1)k})_{\infty}}{(q)_{\infty}} \frac{(q^{a+1+(n-i)k})_{\infty}}{(q^{2a+2b+2+2(n-i)k})_{\infty}} \frac{(q)_{2b+2(n-i)k}\,(q)_{ik}}{(q)_{2b+(n-i)k}\,(q)_k}$$

$$\times \prod_{i=1}^{n} \frac{(q^{2b+2+2(n-i)k};q^2)_b}{(q^{2a+2+2(n-i)k};q^2)_b},$$

which gives the analytic continuation of $_q\mathrm{Pr}_{n,n}(a,b,k)$ as a function of a. We omit the tedious verification that it is an entire function of a. Evaluating the analytic continuation at $a = -1 - 2b - (n-1)k$, we obtain

(10.36)
$$_q\mathrm{Pr}_{n,n}(-1 - 2b - (n-1)k, b, k)$$

$$= \lim_{\epsilon \to 0} \prod_{i=1}^{n} \frac{(q^{\epsilon+1+(n-i)k})_\infty}{(q)_\infty} \frac{(q^{\epsilon-2b-(i-1)k})_\infty}{(q^{2\epsilon-2b-2(i-1)k})_\infty} \frac{(q)_{2b+2(n-i)k}}{(q)_{2b+(n-i)k}} \frac{(q)_{ik}}{(q)_k}$$

$$\times \frac{(q^{2b+2+2(n-i)k}; q^2)_b}{(q^{2\epsilon-4b-2(i-1)k}; q^2)_b}$$

$$= \prod_{i=1}^{n} \frac{1}{(q)_{(n-i)k}} \frac{1}{(q^{-2b-2(i-1)k})_{(i-1)k}} \frac{1}{2} (q^{2b+1+(n-i)k})_{(n-i)k} \frac{(q)_{ik}}{(q)_k}$$

$$\times \frac{(q^{2b+2+2(n-i)k}; q^2)_b}{(q^{-4b-2(i-1)k}; q^2)_b}$$

$$= 2^{-n} \prod_{i=1}^{n} \frac{(q)_{ik}}{(q)_{(i-1)k} (q)_k} \frac{(q^{2b+1+(i-1)k})_{(i-1)k}}{(q^{-2b-2(i-1)k})_{(i-1)k}} \frac{(q^{2b+2+2(i-1)k}; q^2)_b}{(q^{-4b-2(i-1)k}; q^2)_b}.$$

In the last step, we have used the fact that if we run the product $\prod_{i=1}^{n}$ backwards, then $n - i$ becomes $i - 1$. Observe that

(10.37)
$$\prod_{i=1}^{n} \frac{(q)_{ik}}{(q)_{(i-1)k} (q)_k} = \frac{(q)_{nk}}{(q)_k^n}$$

and

(10.38) $$(x)_n = \prod_{i=0}^{n-1} (1 - xq^i) = \prod_{i=0}^{n-1} (-xq^i)(1 - \frac{1}{xq^i}) = (-x)^n q^{\binom{n}{2}} (\frac{q^{1-n}}{x})_n.$$

By (10.38), we have

(10.39) $$\frac{(q^{2b+1+(i-1)k})_{(i-1)k}}{(q^{-2b-2(i-1)k})_{(i-1)k}} = (-1)^{(i-1)k} q^{((2b+1)(i-1)k+(i-1)^2k^2+\binom{(i-1)k}{2})}$$

and, replacing q by q^2, we have

(10.40)
$$\frac{(q^{2b+2+2(i-1)k}; q^2)_b}{(q^{-4b-2(i-1)k}; q^2)_b} = (-1)^b q^{(2b^2+2b+2b(i-1)k+2\binom{b}{2})}$$

$$= (-1)^b q^{(3b^2+b+2b(i-1)k)}.$$

Substituting these results into (10.36), we obtain
(10.41)

$$
{}_q\mathrm{Pr}_{n,n}(-1 - 2b - (n-1)k, b, k)
$$

$$
= 2^{-n} \frac{(q)_{nk}}{(q)_k^n} \prod_{i=1}^{n} (-1)^{(i-1)k} \, q^{\left((2b+1)(i-1)k + (i-1)^2 k^2 + \binom{(i-1)k}{2}\right)}
$$

$$
\times \ (-1)^b q^{\left(3b^2 + b + 2b(i-1)k\right)}
$$

$$
= 2^{-n} \frac{(q)_{nk}}{(q)_k^n} (-1)^{\left(nb + \binom{n}{2}k\right)} \, q^{\left(\sum\limits_{i=1}^{n} 3b^2 + b + (4b+1)(i-1)k + (i-1)^2 k^2 + \binom{(i-1)k}{2}\right)}.
$$

Comparing (10.26) and (10.41), we need only show that the exponents of q are the same. We have

(10.42)
$$
\binom{A+B}{2} = \binom{A}{2} + AB + \binom{B}{2}
$$

and

(10.43)
$$
\binom{AB}{2} = A\binom{B}{2} + \binom{A}{2}B^2.
$$

The exponent of q in (10.26) is
(10.44)

$$
nb^2 + \binom{n}{2}\binom{k}{2} + n\binom{2b+1+(n-1)k}{2}
$$

$$
= nb^2 + \binom{n}{2}\binom{k}{2} + n\left[\binom{2b+1}{2} + (2b+1)(n-1)k + \binom{(n-1)k}{2}\right]
$$

$$
= n\left[b^2 + \binom{2b+1}{2}\right] + \binom{n}{2}\binom{k}{2} + 2(2b+1)\binom{n}{2}k
$$

$$
+ n\left[\binom{n-1}{2}k^2 + (n-1)\binom{k}{2}\right]
$$

$$
= n(3b^2 + b) + 3\binom{n}{2}\binom{k}{2} + (4b+2)\binom{n}{2}k + 3\binom{n}{3}k^2.
$$

For fixed $r \geq 1$, we have

(10.45)
$$
\sum_{i=1}^{n} \binom{i-1}{r-1} = \binom{n}{r}.
$$

The exponent of q in (10.41) is
(10.46)

$$\sum_{i=1}^{n} 3b^2 + b + (4b+1)(i-1)k + (i-1)^2 k^2 + \binom{(i-1)k}{2}$$

$$= n(3b^2 + b) + (4b+1)\binom{n}{2}k + \sum_{i=1}^{n} \Big((i-1) + (i-1)(i-2)\Big)k^2$$

$$+ (i-1)\binom{k}{2} + \binom{i-1}{2}k^2$$

$$= n(3b^2 + b) + (4b+1)\binom{n}{2}k + \binom{n}{2}k^2 + \binom{n}{2}\binom{k}{2} + 3\sum_{i=1}^{n}\binom{i-1}{2}k^2$$

$$= n(3b^2 + b) + (4b+2)\binom{n}{2}k + 3\binom{n}{2}\binom{k}{2} + 3\binom{n}{3}k^2.$$

The proof of Theorem 4 (1.18) is now complete since (10.44) and (10.46) are equal.

11. The parameter r

In this section, we prove Lemma 23 by which we may evaluate $_qBC_{n,m,r}(a,b,k)$.

Lemma 23. *Let* $r \geq s \geq 0$. *Then*

$$
q^{\left(sa+s(n-m-r)k+\binom{s}{2}k\right)} {}_qBC_{n,m,r}(a,b,k)
$$
(11.1)
$$
= \sum_{i=0}^{s} (-1)^i q^{\binom{i}{2}k} \begin{bmatrix} s \\ i \end{bmatrix}_{q^k} {}_qBC_{n,m+i,r-s}(a,b,k).
$$

Proof. We proceed by induction on s. The $s = 0$ case is trivial. Multiplying (8.12) by $q^{a+(n-m-r)k}$ gives

$$
q^{a+(n-m-r)k} {}_qBC_{n,m,r}(a,b,k)
$$
(11.2)
$$
= {}_qBC_{n,m,r-1}(a,b,k) - {}_qBC_{n,m+1,r-1}(a,b,k), \quad r \geq 1,
$$

which is the $s = 1$ case. Assume that (11.1) holds for a given s where $s \geq 1$. By (11.2), we have

$$
q^{a+(n-m-i-r+s)k} {}_qBC_{n,m+i,r-s}(a,b,k)
$$
(11.3)
$$
= {}_qBC_{n,m+i,r-s-1}(a,b,k) - {}_qBC_{n,m+i+1,r-s-1}(a,b,k), \quad r \geq s+1.
$$

Multiplying (11.1) by $q^{a+(n-m-r+s)k}$, using (11.3), and rearranging, we have

$$
q^{\left((s+1)a+(s+1)(n-m-r)k+\binom{s+1}{2}k\right)} {}_qBC_{n,m,r}(a,b,k)
$$

$$
= \sum_{i=0}^{s} (-1)^i q^{\left(a+\binom{i}{2}k+(n-m-r+s)k\right)} \begin{bmatrix} s \\ i \end{bmatrix}_{q^k} {}_qBC_{n,m+i,r-s}(a,b,k),
$$

$$
= \sum_{i=0}^{s} (-1)^i q^{\binom{i+1}{2}k} \begin{bmatrix} s \\ i \end{bmatrix}_{q^k}
$$
(11.4)
$$
\times \left({}_qBC_{n,m+i,r-s-1}(a,b,k) - {}_qBC_{n,m+i+1,r-s-1}(a,b,k) \right)
$$

$$
= \sum_{i=0}^{s} (-1)^i q^{\binom{i+1}{2}k} \begin{bmatrix} s \\ i \end{bmatrix}_{q^k} {}_qBC_{n,m+i,r-s-1}(a,b,k)
$$

$$
+ \sum_{i=1}^{s+1} (-1)^i q^{\binom{i}{2}k} \begin{bmatrix} s \\ i-1 \end{bmatrix}_{q^k} {}_qBC_{n,m+i,r-s-1}(a,b,k), \quad r \geq s+1.
$$

Observe that

$$
q^{ik} \begin{bmatrix} s \\ i \end{bmatrix}_{q^k} + \begin{bmatrix} s \\ i-1 \end{bmatrix}_{q^k} = \left(q^{ik} \frac{(1-q^{(s+1-i)k})}{(1-q^{(s+1)k})} + \frac{(1-q^{ik})}{(1-q^{(s+1)k})} \right) \begin{bmatrix} s+1 \\ i \end{bmatrix}_{q^k}
$$
(11.5)
$$
= \begin{bmatrix} s+1 \\ i \end{bmatrix}_{q^k}.
$$

Since the two additional terms are both zero, we may run both of the sums on the extreme right side of (11.4) from 0 to $s + 1$. This gives

$$q^{\left((s+1)a+(s+1)(n-m-r)k+\binom{s+1}{2}k\right)} {}_q BC_{n,m,r}(a,b,k)$$

(11.6)

$$= \sum_{i=0}^{s+1} (-1)^i q^{\binom{i}{2}k} \left(q^{ik} \begin{bmatrix} s \\ i \end{bmatrix}_{q^k} + \begin{bmatrix} s \\ i-1 \end{bmatrix}_{q^k} \right) {}_q BC_{n,m+i,r-s-1}(a,b,k)$$

$$= \sum_{i=0}^{s+1} (-1)^i q^{\binom{i}{2}k} \begin{bmatrix} s+1 \\ i \end{bmatrix}_{q^k} {}_q BC_{n,m+i,r-s-1}(a,b,k), \quad r \geq s+1.$$

Our induction is complete since (11.6) is (11.1) with s replaced by $s+1$. $\square$

We may evaluate ${}_q BC_{n,m,r}(a,b,k)$ by taking $s = r$ in Lemma 23 (11.1) and using Theorem 4 (1.18).

12. The q-Macdonald-Morris conjecture for B_n, $B_n^\vee$, C_n, $C_n^\vee$ and D_n

In this section, we discuss the q-Macdonald-Morris conjecture for B_n, $B_n^\vee$, C_n, $C_n^\vee$ and D_n. We observe that the B_n, $B_n^\vee$ and D_n cases of the conjecture follow from Theorem 4 (1.18). We give some of the details for C_n and $C_n^\vee$ and acknowledge the priority of Gustafson's recent proof [Gu1] of these cases of the conjecture. This illustrates the basic steps required to apply our methods to the conjecture when the reduced irreducible root system R does not have a minuscule weight.

Following Morris [Mo1, Appendix C], we associate the functions

$$(12.1) \quad {}_qb_n(a,k;t_1,\ldots,t_n) = \prod_{i=1}^{n}(t_i)_a(\frac{q}{t_i})_a \prod_{1\le i<j\le n}(\frac{t_i}{t_j})_k(q\frac{t_j}{t_i})_k(t_it_j)_k(\frac{q}{t_it_j})_k$$

$$(12.2)$$
$$
{}_qb_n^\vee(b,k;t_1,\ldots,t_n) = \prod_{i=1}^{n}(t_i^2;q^2)_b(\frac{q^2}{t_i^2};q^2)_b \prod_{1\le i<j\le n}(\frac{t_i}{t_j})_k(q\frac{t_j}{t_i})_k(t_it_j)_k(\frac{q}{t_it_j})_k
$$

$$(12.3) \quad {}_qc_n(b,k;t_1,\ldots,t_n) = \prod_{i=1}^{n}(t_i^2)_b(\frac{q}{t_i^2})_b \prod_{1\le i<j\le n}(\frac{t_i}{t_j})_k(q\frac{t_j}{t_i})_k(t_it_j)_k(\frac{q}{t_it_j})_k$$

$$(12.4)$$
$$
{}_qc_n^\vee(a,k;t_1,\ldots,t_n)
$$
$$
= \prod_{i=1}^{n}(t_i)_a(\frac{q}{t_i})_a \prod_{1\le i<j\le n}(\frac{t_i}{t_j};q^2)_k(q^2\frac{t_j}{t_i};q^2)_k(t_it_j;q^2)_k(\frac{q^2}{t_it_j};q^2)_k
$$

with the named root systems. As with BC_n, we use capital letters to denote the constant term in the function associated with the root system. Thus

$$
{}_qB_n(a,k) = [1]\,{}_qb_n(a,k;t_1,\ldots,t_n),
$$
$$
{}_qB_n^\vee(b,k) = [1]\,{}_qb_n^\vee(b,k;t_1,\ldots,t_n),
$$
$$(12.5) \qquad {}_qC_n(b,k) = [1]\,{}_qc_n(b,k;t_1,\ldots,t_n),$$
$$
{}_qC_n^\vee(a,k) = [1]\,{}_qc_n^\vee(a,k;t_1,\ldots,t_n),
$$
$$
{}_qD_n(k) = [1]\,{}_qd_n(k;t_1,\ldots,t_n).
$$

Observe that

$$
{}_qb_n(a,k;t_1,\ldots,t_n) = {}_qbc_n(a,0,k;t_1,\ldots,t_n),
$$
$$(12.6) \qquad {}_qb_n^\vee(b,k;t_1,\ldots,t_n) = {}_qbc_n(0,b,k;\frac{\sqrt{q}}{t_n},\ldots,\frac{\sqrt{q}}{t_1}),$$
$$
{}_qd_n(k;t_1,\ldots,t_n) = {}_qbc_n(0,0,k;t_1,\ldots,t_n).
$$

Since the constant term in a polynomial is fixed by the substitution $t_i \leftrightarrow \sqrt{q}/t_{n-i+1}$,

$1 \le i \le n$, we obtain

$$_qB_n(a,k) = {}_qBC_n(a,0,k),$$

(12.7) $$_qB_n^{\vee}(b,k) = {}_qBC_n(0,b,k),$$

$$_qD_n(k) = {}_qBC_n(0,0,k).$$

Thus the B_n, $B_n^{\vee}$ and D_n cases of the q-Macdonald-Morris conjecture follow from Theorem 4 (1.18).

Since D_n is a subsystem of each of the root systems B_n, $B_n^{\vee}$, C_n and $C_n^{\vee}$, we may use the geometry of the simple roots of D_n in each case. We may explicitly express the geometry by setting $a = b = 0$ in Lemma 6 (3.58) and (3.59). This gives

$$_qd_n(k; t_2, \ldots, t_{v-1}, t_1, t_v, \ldots, t_n)$$

(12.8)
$$= \prod_{j=2}^{v-1} (q\frac{t_1}{t_j})_k (\frac{t_j}{t_1})_k (t_1 t_j)_k (\frac{q}{t_1 t_j})_k \prod_{j=v}^{n} (\frac{t_1}{t_j})_k (q\frac{t_j}{t_1})_k (t_1 t_j)_k (\frac{q}{t_1 t_j})_k$$

$$\times \prod_{2 \le i < j \le n} (\frac{t_i}{t_j})_k (q\frac{t_j}{t_i})_k (t_i t_j)_k (\frac{q}{t_i t_j})_k, \quad 2 \le v \le n+1,$$

and

$$_qd_n(k; t_2, \ldots, t_v, \frac{1}{t_1}, t_{v+1}, \ldots, t_n)$$

(12.9)
$$= \prod_{j=2}^{v} (q\frac{t_1}{t_j})_k (\frac{t_j}{t_1})_k (t_1 t_j)_k (\frac{q}{t_1 t_j})_k \prod_{j=v+1}^{n} (q\frac{t_1}{t_j})_k (\frac{t_j}{t_1})_k (q t_1 t_j)_k (\frac{1}{t_1 t_j})_k$$

$$\times \prod_{2 \le i < j \le n} (\frac{t_i}{t_j})_k (q\frac{t_j}{t_i})_k (t_i t_j)_k (\frac{q}{t_i t_j})_k, \quad 1 \le v \le n.$$

When we try to apply the q-engine (4.4) of our q-machine to the root systems C_n and $C_n^{\vee}$, we encounter essentially the same problem. We wish to compute $t_1 \partial_q / \partial_q t_1 {}_qc_n(b, k; t_1, \ldots, t_n)$ and $t_1 \partial_{q^2} / \partial_{q^2} t_1 {}_qc_n^{\vee}(a, k; t_1, \ldots, t_n)$. This requires the substitutions $t_1 \to q t_1$ and $t_1 \to q^2 t_1$, respectively. Observe that

(12.10) $$(t_1^2)_b (\frac{q}{t_1^2})_b \to (q^2 t_1^2)_b (\frac{1}{q t_1^2})_b \text{ under } t_1 \to q t_1$$

and

(12.11) $$(t_1)_a (\frac{q}{t_1})_a \to (q^2 t_1)_a (\frac{1}{q t_1})_a \text{ under } t_1 \to q^2 t_1.$$

Thus, the zeros of $(t_1^2)_b(q/t_1^2)_b$ and $(t_1)_a(q/t_1)_a$ are shifted by 2 in the q-differen-tiation process. Observe that

$$(12.12) \qquad (t_1^2)_b(\frac{q}{t_1^2})_b \leftrightarrow (qt_1^2)_b(\frac{1}{t_1^2})_b \text{ under } t_1 \leftrightarrow \frac{1}{t_1}$$

and

$$(12.13) \qquad (t_1)_a(\frac{q}{t_1})_a \leftrightarrow (qt_1)_a(\frac{1}{t_1})_a \text{ under } t_1 \leftrightarrow \frac{1}{t_1}.$$

Thus, the zeros of $(t_1^2)_b(q/t_1^2)_b$ and $(t_1)_a(q/t_1)_a$ are only shifted by 1 by the sub-stitution $t_1 \leftrightarrow 1/t_1$, which we are prepared to use in conjunction with (12.8) and (12.9). The fact that this difficulty also occurs for G_2, $G_2^\vee$, F_4 and $F_4^\vee$ did not stop Habsieger [Ha1], Zeilberger [Ze1, Ze2] or Garvan and Gonnet [GG1] from treating the q-Macdonald-Morris conjecture for these root systems. We may remedy this difficulty for C_n and $C_n^\vee$ by breaking the appropriate partial q-derivatives into 2 parts as follows. Using (4.6), we have

$$
\begin{aligned}
(12.14) \quad t_1 \frac{\partial_q}{\partial_q t_1}\left((t_1^2)_b(\frac{q}{t_1^2})_b\right) &= \frac{1}{(1-q)}\left((t_1^2)_b(\frac{q}{t_1^2})_b - (q^2 t_1^2)_b(\frac{1}{qt_1^2})_b\right) \\
&= \frac{1}{(1-q)}\left((t_1^2)_b(\frac{q}{t_1^2})_b - (qt_1^2)_b(\frac{1}{t_1^2})_b\right) \\
&\quad + \frac{1}{(1-q)}\left((qt_1^2)_b(\frac{1}{t_1^2})_b - (q^2 t_1^2)_b(\frac{1}{qt_1^2})_b\right) \\
&= \frac{(1-q^b)}{(1-q)}\frac{(t_1^2+1)}{(t_1^2-q^b)}(t_1^2)_b(\frac{q}{t_1^2})_b + \frac{(1-q^b)}{(1-q)}\frac{(qt_1^2+1)}{(qt_1^2-q^b)}(qt_1^2)_b(\frac{1}{t_1^2})_b
\end{aligned}
$$

and
$$
\begin{aligned}
(12.15) \\
t_1 \frac{\partial_{q^2}}{\partial_{q^2} t_1}\left((t_1)_a(\frac{q}{t_1})_a\right) &= \frac{1}{(1-q^2)}\left((t_1)_a(\frac{q}{t_1})_a - (q^2 t_1)_a(\frac{1}{qt_1})_a\right) \\
&= \frac{1}{(1-q^2)}\left((t_1)_a(\frac{q}{t_1})_a - (qt_1)_a(\frac{1}{t_1})_a\right) \\
&\quad + \frac{1}{(1-q^2)}\left((qt_1)_a(\frac{1}{t_1})_a - (q^2 t_1)_a(\frac{1}{qt_1})_a\right) \\
&= \frac{(1-q^a)}{(1-q^2)}\frac{(t_1+1)}{(t_1-q^a)}(t_1)_a(\frac{q}{t_1})_a + \frac{(1-q^a)}{(1-q^2)}\frac{(qt_1+1)}{(qt_1-q^a)}(qt_1)_a(\frac{1}{t_1})_a.
\end{aligned}
$$

By the product rule (4.2) for q-derivatives, we have

$$t_1 \frac{\partial_q}{\partial_q t_1} \left(f(t_1, \dots, t_n) \, (t_1^2)_b (\frac{q}{t_1^2})_b \, g(t_1, \dots, t_n) \right)$$

$$= t_1 \frac{\partial_q}{\partial_q t_1} \left(f(t_1, \dots, t_n) \right) (t_1^2)_b (\frac{q}{t_1^2})_b \, g(t_1, \dots, t_n)$$

(12.16)
$$+ f(qt_1, \dots, t_n) \frac{(1-q^b)}{(1-q)} \frac{(t_1^2+1)}{(t_1^2-q^b)} (t_1^2)_b (\frac{q}{t_1^2})_b \, g(t_1, \dots, t_n)$$

$$+ f(qt_1, \dots, t_n) \, (qt_1^2)_b (\frac{1}{t_1^2})_b \, t_1 \frac{\partial_q}{\partial_q t_1} \left(g(t_1, \dots, t_n) \right)$$

$$+ f(qt_1, \dots, t_n) \frac{(1-q^b)}{(1-q)} \frac{(qt_1^2+1)}{(qt_1^2-q^b)} (qt_1^2)_b (\frac{1}{t_1^2})_b \, g(qt_1, \dots, t_n)$$

and

$$t_1 \frac{\partial_{q^2}}{\partial_{q^2} t_1} \left(f(t_1, \dots, t_n) \, (t_1)_a (\frac{q}{t_1})_a \, g(t_1, \dots, t_n) \right)$$

$$= t_1 \frac{\partial_{q^2}}{\partial_{q^2} t_1} \left(f(t_1, \dots, t_n) \right) (t_1)_a (\frac{q}{t_1})_a \, g(t_1, \dots, t_n)$$

(12.17)
$$+ f(q^2 t_1, \dots, t_n) \frac{(1-q^a)}{(1-q^2)} \frac{(t_1+1)}{(t_1-q^a)} (t_1)_a (\frac{q}{t_1})_a \, g(t_1, \dots, t_n)$$

$$+ f(q^2 t_1, \dots, t_n) \, (qt_1)_a (\frac{1}{t_1})_a \, t_1 \frac{\partial_{q^2}}{\partial_{q^2} t_1} \left(g(t_1, \dots, t_n) \right)$$

$$+ f(q^2 t_1, \dots, t_n) \frac{(1-q^a)}{(1-q^2)} \frac{(qt_1+1)}{(qt_1-q^a)} (qt_1)_a (\frac{1}{t_1})_a \, g(q^2 t_1, \dots, t_n).$$

The idea is to q-differentiate the factors associated with the roots $e_1 - e_v$ with v running from 2 to n, followed by the first part of (12.14) or (12.15). Then we q-differentiate the factors associated with the roots $e_1 + e_v$ with v running from n down to 2, followed by the second part of (12.14) or (12.15).

Taking

$$f(t_1, \dots, t_n) = \prod_{i=2}^{n} (t_i^2)_b (\frac{q}{t_i^2})_b \prod_{1 \le i < j \le n} (\frac{t_i}{t_j})_k (q \frac{t_j}{t_i})_k,$$

(12.18)

$$g(t_1, \dots, t_n) = \prod_{1 \le i < j \le n} (t_i t_j)_k (\frac{q}{t_i t_j})_k,$$

in (12.16) and using the geometry (12.8) and (12.9) of the simple roots of D_n, we obtain

$$
t_1 \frac{\partial_q}{\partial_q t_1} \left({}_q c_n(b,k;t_1,\ldots,t_n) \right)
$$

$$
= \frac{(1-q^k)}{(1-q)} \sum_{v=2}^n \frac{(t_1+t_v)}{(t_1-q^k t_v)} \, {}_q c_n(b,k;t_2,\ldots,t_{v-1},t_1,t_v,\ldots,t_n)
$$

(12.19)
$$
+ \frac{(1-q^b)}{(1-q)} \frac{(t_1^2+1)}{(t_1^2-q^b)} \, {}_q c_n(b,k;t_2,\ldots,t_n,t_1)
$$

$$
+ \frac{(1-q^k)}{(1-q)} \sum_{v=2}^n \frac{(t_1 t_v+1)}{(t_1 t_v-q^k)} \, {}_q c_n(b,k;t_2,\ldots,t_v,\frac{1}{t_1},t_{v+1},\ldots,t_n)
$$

$$
+ \frac{(1-q^b)}{(1-q)} \frac{(q t_1^2+1)}{(q t_1^2-q^b)} \, {}_q c_n(b,k;\frac{1}{t_1},t_2,\ldots,t_n).
$$

Taking

$$
f(t_1,\ldots,t_n) = \prod_{i=2}^n (t_i)_a (\frac{q}{t_i})_a \prod_{1\le i<j\le n} (\frac{t_i}{t_j};q^2)_k (q^2 \frac{t_j}{t_i};q^2)_k,
$$

(12.20)
$$
g(t_1,\ldots,t_n) = \prod_{1\le i<j\le n} (t_i t_j;q^2)_k (\frac{q^2}{t_i t_j};q^2)_k,
$$

in (12.17) and using the geometry (12.8) and (12.9) of the simple roots of D_n with q replaced by q^2, we obtain

$$
t_1 \frac{\partial_{q^2}}{\partial_{q^2} t_1} \left({}_q c_n^\vee(a,k;t_1,\ldots,t_n) \right)
$$

$$
= \frac{(1-q^{2k})}{(1-q^2)} \sum_{v=2}^n \frac{(t_1+t_v)}{(t_1-q^{2k} t_v)} \, {}_q c_n^\vee(a,k;t_2,\ldots,t_{v-1},t_1,t_v,\ldots,t_n)
$$

(12.21)
$$
+ \frac{(1-q^a)}{(1-q^2)} \frac{(t_1+1)}{(t_1-q^a)} \, {}_q c_n^\vee(a,k;t_2,\ldots,t_n,t_1)
$$

$$
+ \frac{(1-q^{2k})}{(1-q^2)} \sum_{v=2}^n \frac{(t_1 t_v+1)}{(t_1 t_v-q^{2k})} \, {}_q c_n^\vee(a,k;t_2,\ldots,t_v,\frac{1}{t_1},t_{v+1},\ldots,t_n)
$$

$$
+ \frac{(1-q^a)}{(1-q^2)} \frac{(q t_1+1)}{(q t_1-q^a)} \, {}_q c_n^\vee(a,k;\frac{1}{t_1},t_2,\ldots,t_n).
$$

We now turn to the antisymmetries of the terms occurring in the partial q-derivatives (12.19) and (12.21). For each term which is associated with D_n, that is which involves the parameter k, the analysis proceeds as before. For both (12.19) and (12.21), the terms occurring in the sums of the form $\sum_{v=2}^n$ on the right side are antisymmetric under the substitutions

(12.22)
$$
t_1 \leftrightarrow t_v, \text{ and } t_1 \leftrightarrow \frac{1}{t_v}, \ 2 \le v \le n,
$$

respectively. We have

$$\frac{(1-q^b)}{(1-q)}\frac{(t_1^2+1)}{(t_1^2-q^b)}\,_q c_n(b,k;t_2,\ldots,t_n,t_1)$$

$$=\,_q h(b;t_1^2,1)\prod_{i=2}^{n}(t_i^2)_b(\frac{q}{t_i^2})_b\,_q d_n(k;t_2,\ldots,t_n,t_1),$$

(12.23)

$$\frac{(1-q^b)}{(1-q)}\frac{(qt_1^2+1)}{(qt_1^2-q^b)}\,_q c_n(b,k;\frac{1}{t_1},t_2,\ldots,t_n)$$

$$=\,_q h(b;qt_1^2,1)\prod_{i=2}^{n}(t_i^2)_b(\frac{q}{t_i^2})_b\,_q d_n(k;\frac{1}{t_1},t_2,\ldots,t_n),$$

and

$$\frac{(1-q^a)}{(1-q^2)}\frac{(t_1+1)}{(t_1-q^a)}\,_q c_n^{\vee}(a,k;t_2,\ldots,t_n,t_1)$$

$$=\frac{(1-q)}{(1-q^2)}\,_q h(a;t_1,1)\prod_{i=2}^{n}(t_i)_a(\frac{q}{t_i})_a\,_{q^2} d_n(k;t_2,\ldots,t_n,t_1),$$

(12.24)

$$\frac{(1-q^a)}{(1-q^2)}\frac{(qt_1+1)}{(qt_1-q^a)}\,_q c_n^{\vee}(a,k;\frac{1}{t_1},t_2,\ldots,t_n)$$

$$=\frac{(1-q)}{(1-q^2)}\,_q h(a;qt_1,1)\prod_{i=2}^{n}(t_i)_a(\frac{q}{t_i})_a\,_{q^2} d_n(k;\frac{1}{t_1},t_2,\ldots,t_n).$$

By the antisymmetry (4.18) of $_q h(k;s,t)$ under $s,t \leftrightarrow 1/s,1/t$, and the symmetries (3.35) of $_q d_n(k;t_1,\ldots,t_n)$ under $t_n \leftrightarrow 1/t_n$ and $t_1 \leftrightarrow q/t_1$, we see that the terms occurring in the second and fourth expressions on the right side of the partial q-derivative (12.19) are antisymmetric under the substitutions

(12.25) $$t_1 \leftrightarrow \frac{1}{t_1} \text{ and } t_1 \leftrightarrow \frac{1}{qt_1},$$

respectively. Similarly, replacing q by q^2, we see that the terms occurring in the second and fourth expressions on the right side of the partial q-derivative (12.21) are antisymmetric under the substitutions

(12.26) $$t_1 \leftrightarrow \frac{1}{t_1} \text{ and } t_1 \leftrightarrow \frac{1}{q^2 t_1},$$

respectively. The antisymmetries of (12.25) and (12.26) also follow by inspection of the second expression on the right side of (12.14) and (12.15), respectively.

Thus Lemmas 10, 12 and 14 may be explicitly expressed in terms of $_q c_n(b,k; t_1,\ldots,t_n)$ and $_q c_n^{\vee}(a,k;t_1,\ldots,t_n)$. Lemma 10 (6.12) gives

(12.27)
$$[1]\,t_v\,\omega(t_1,\ldots,t_n)\,_q c_n(b,k;t_1,\ldots,t_n)$$

$$=q^k\,[1]\,t_{v-1}\,\omega(t_1,\ldots,t_n)\,_q c_n(b,k;t_1,\ldots,t_n),$$

(12.28)
$$[1]\, t_v\, \omega(t_1,\dots,t_n)\, {}_q c_n^{\vee}(b,k;t_1,\dots,t_n)$$
$$= q^{2k}\, [1]\, t_{v-1}\, \omega(t_1,\dots,t_n)\, {}_q c_n^{\vee}(b,k;t_1,\dots,t_n),$$

$$\omega(t_1,\dots,t_n) = \omega(t_1,\dots,t_{v-2},t_v,t_{v-1},t_{v+1},\dots,t_n),\ 2 \le v \le n.$$

Lemma 12 (6.19) gives
(12.29)
$$[1]\, \frac{1}{t_n}\, \omega(t_1,\dots,t_n)\, {}_q c_n(b,k;t_1,\dots,t_n)$$
$$= q^b\, [1]\, t_n\, \omega(t_1,\dots,t_n)\, {}_q c_n(b,k;t_1,\dots,t_n),$$

(12.30)
$$[1]\, (1+\frac{1}{t_n})\, \omega(t_1,\dots,t_n)\, {}_q c_n^{\vee}(a,k;t_1,\dots,t_n)$$
$$= q^a\, [1]\, (1+t_n)\, \omega(t_1,\dots,t_n)\, {}_q c_n^{\vee}(a,k;t_1,\dots,t_n),$$

$$\omega(t_1,\dots,t_n) = \omega(t_1,\dots,t_{n-1},\frac{1}{t_n}).$$

Lemma 14 (6.31) gives

(12.31)
$$[1]\, t_1\, \omega(t_1,\dots,t_n)\, {}_q c_n(b,k;t_1,\dots,t_n)$$
$$= q^{b+1}\, [1]\, \frac{1}{t_1}\, \omega(t_1,\dots,t_n)\, {}_q c_n(b,k;t_1,\dots,t_n),$$
$$\omega(t_1,\dots,t_n) = \omega(\frac{q}{t_1},t_2,\dots,t_n),$$

and

(12.32)
$$[1]\, (q+t_1)\, \omega(t_1,\dots,t_n)\, {}_q c_n^{\vee}(a,k;t_1,\dots,t_n)$$
$$= q^{a+1}\, [1]\, (1+\frac{q}{t_1})\, \omega(t_1,\dots,t_n)\, {}_q c_n^{\vee}(a,k;t_1,\dots,t_n),$$
$$\omega(t_1,\dots,t_n) = \omega(\frac{q^2}{t_1},t_2,\dots,t_n).$$

For the results (12.29) and (12.31), which are related to ${}_q c_n(b,k;t_1,\dots,t_n)$, we may use the type of rearrangements which are used in the proof of Lemma 22, especially (8.22).

The reader may ask why we do not proceed to establish the q-Macdonald-Morris conjecture for C_n and $C_n^{\vee}$ by using the partial q-derivative identities (12.19) and (12.21) and the q-transportation theory (12.27)–(12.32). This is to acknowledge the priority of Gustafson's recent proof [Gu1] of these cases of the conjecture.

13. Conclusion

We give [Ka4] a simple proof of an Aomoto type extension of the q-Morris theorem. Gustafson [Gu1] incorporates the Askey-Wilson integral [AW1] into an elegant multivariable q-Selberg integral which gives the q-Macdonald-Morris conjecture for B_n, $B_n^\vee$, C_n, $C_n^\vee$, D_n and BC_n. We give [Ka5] a simple proof of an Aomoto type extension of Gustafson's theorem.

Let R be a reduced, irreducible root system of an isolated type. We may give an identity for a partial q-derivative which allows us to use the q-engine (4.4) of our q-machine. If R does not have a minuscule weight, then it has a quasi-minuscule weight and we may follow the approach given in Section 12 for C_n and $C_n^\vee$. The recent work of Garvan [Ga1, Ga2] and Garvan and Gonnet [GG1] suggests that there exist Aomoto type extensions of the q-Macdonald-Morris conjecture for R. We should be able to use our q-machine to prove the required functional equations. Following [Ka4, Ka5], we should be able to give a simple proof, which is read off from the identity for a partial q-derivative. The author has recently carried out these calculations by hand for G_2.

Hopefully there is a general formula for an Aomoto type extension of the q-Macdonald-Morris conjecture for R which can be expressed in terms of the geometry or algebra of R and a classification free proof is not far behind.

Acknowledgment

I am grateful to Ian Goulden who listened to the earliest stages of this paper and provided constant encouragement. I also thank the Department of Combinatorics and Optimization at the University of Waterloo in Ontario, Canada for providing an excellent environment during my visiting appointment. I also thank the referee for many suggestions which improved the presentation.

REFERENCES

[An1] G. E. ANDREWS, *The Theory of Partitions*, Addison-Wesley, Reading, MA, 1976.

[An2] G. E. ANDREWS, *q-Series: Their Development and Applications in Analysis, Number Theory, Combinatorics, Physics, and Computer Algebra*, CBMS Regional Conference Series in Mathematics, number 66, Amer. Math. Soc., Providence, RI, 1986.

[Ao1] K. AOMOTO, *Jacobi polynomials associated with Selberg's integral*, SIAM J. Math. Analysis **18** (1987), 545–549.

[As1] R. ASKEY, *Some basic hypergeometric extensions of integrals of Selberg and Andrews*, SIAM J. Math. Analysis **11** (1980), 938–951.

[As2] R. ASKEY, *Integration and Computers*, to appear in the proceedings of a computer algebra conference edited by G. and D. Chudnovsky.

[AW1] R. A. ASKEY AND J. A. WILSON, *Some basic hypergeometric orthogonal polynomials that generalize Jacobi polynomials*, Memoirs of the AMS, number 319 (1985), American Mathematical Society, Providence, RI.

[Ca1] R. W. CARTER, *Simple Groups of Lie Type*, John Wiley, London, New York, 1972.

[Ga1] F. G. GARVAN, *A beta integral associated with the root system G_2*, SIAM J. Math. Analysis **19** (1989), 1462–1474.

[Ga2] F. G. GARVAN, *A proof of the Macdonald-Morris root system conjecture for F_4*, SIAM J. Math. Analysis **21** (1989), 803–821.

[GG1] F. G. GARVAN AND G. H. GONNET, *A proof of the two parameter q-case of the Macdonald-Morris constant term root system conjecture for $S(F_4)$ and $S(F_4)^\vee$ via Zeilberger's method*, J. Symbolic Comp., to appear.

[GB1] L. C. GROVE AND C. T. BENSON, *Finite Reflection Groups*, second edition, Springer-Verlag, New York, 1985.

[Gu1] R. A. GUSTAFSON, *A generalization of Selberg's beta integral*, Bull. Amer. Math. Soc. **22** (1990), 97–105.

[Ha1] L. HABSIEGER, *La q-conjecture de Macdonald-Morris pour G_2*, C. R. Acad. Sci. **303** (1986), 211–213.

[Ha2] L. HABSIEGER, *Une q-intégrale de Selberg-Askey*, SIAM J. Math. Analysis **19** (1988), 1475–1489.

[He1] G. J. HECKMAN, *Root systems and hypergeometric functions II*, Compositio Math. **64** (1987), 353–373.

[HO1] G. J. HECKMAN AND E. M. OPDAM, *Root systems and hypergeometric functions I*, Compositio Math. **64** (1987), 329–352.

[Hu1] J. E. HUMPHREYS, *Introduction to Lie Algebras and Representation Theory*, Springer-Verlag, New York, 1972.

[Ka1] K. W. J. KADELL, *A proof of Andrews' q-Dyson conjecture for $n = 4$*, Trans. Amer. Math. Soc. **290** (1985), 127–144.

[Ka2] K. W. J. KADELL, *A proof of some q-analogues of Selberg's integral for $k = 1$*, SIAM J. Math. Analysis **19** (1988), 944–968.

[Ka3] K. W. J. KADELL, *A proof of Askey's conjectured q-analogue of Selberg's integral and a conjecture of Morris*, SIAM J. Math. Analysis **19** (1988), 969–986.

[Ka4] K. W. J. KADELL, *A simple proof of an Aomoto type extension of the q-Morris theorem*, Proceedings of the Rademacher Centenary Conference, to appear.

[Ka5] K. W. J. KADELL, *A simple proof of an Aomoto type extension of Gustafson's Askey-Wilson q-Selberg integral*, preprint.

[Ma1] I. G. MACDONALD, *Some conjectures for root systems and finite reflection groups*, SIAM J. Math. Analysis **13** (1982), 988–1007.

[Mo1] W. G. MORRIS, II, *Constant term identities for finite and affine root systems: conjectures and theorems*, Ph.D. dissertation, University of Wisconsin-Madison, January 1982.

[Op1] E. M. OPDAM, *Root systems and hypergeometric functions III*, Compositio Math. **67** (1988), 21–49.

[Op2] E. M. OPDAM, *Root systems and hypergeometric functions IV*, Compositio Math. **67** (1988), 191–209.

[Op3] E. M. OPDAM, *Some applications of hypergeometric shift operators*, Invent. Math. **98** (1989), 1–18.

[Se1] A. SELBERG, *Bemerkninger om et multipelt integral*, Norsk. Mat. Tiddskr. **26** (1944), 71–78.

[St1] J. STEMBRIDGE, *A short proof of Macdonald's conjecture for the root systems of type A*, Proc. Amer. Math. Soc. **102** (1988), 777–786.

[Ze1] D. ZEILBERGER, *A proof of the G_2 case of Macdonald's root system-Dyson conjecture*, SIAM J. Math. Analysis **18** (1987), 880–883.

[Ze2] D. ZEILBERGER, *A unified approach to Macdonald's root-system conjectures*, SIAM J. Math. Analysis **19** (1988), 987–1013.

[Ze3] D. ZEILBERGER, *A Stembridge-Stanton style elementary proof of the Habsieger-Kadell q-Morris identity*, Discrete Math. **79**, (1989/90), 313–322.

Editorial Information

To be published in the *Memoirs*, a paper must be correct, new, nontrivial, and significant. Further, it must be well written and of interest to a substantial number of mathematicians. Piecemeal results, such as an inconclusive step toward an unproved major theorem or a minor variation on a known result, are in general not acceptable for publication. *Transactions* Editors shall solicit and encourage publication of worthy papers. Papers appearing in *Memoirs* are generally longer than those appearing in *Transactions* with which it shares an editorial committee.

As of January 6, 1994, the backlog for this journal was approximately 7 volumes. This estimate is the result of dividing the number of manuscripts for this journal in the Providence office that have not yet gone to the printer on the above date by the average number of monographs per volume over the previous twelve months, reduced by the number of issues published in four months (the time necessary for preparing an issue for the printer). (There are 6 volumes per year, each containing at least 4 numbers.)

A Copyright Transfer Agreement is required before a paper will be published in this journal. By submitting a paper to this journal, authors certify that the manuscript has not been submitted to nor is it under consideration for publication by another journal, conference proceedings, or similar publication.

Information for Authors and Editors

Memoirs are printed by photo-offset from camera copy fully prepared by the author. This means that the finished book will look exactly like the copy submitted.

The paper must contain a *descriptive title* and an *abstract* that summarizes the article in language suitable for workers in the general field (algebra, analysis, etc.). The *descriptive title* should be short, but informative; useless or vague phrases such as "some remarks about" or "concerning" should be avoided. The *abstract* should be at least one complete sentence, and at most 300 words. Included with the footnotes to the paper, there should be the 1991 *Mathematics Subject Classification* representing the primary and secondary subjects of the article. This may be followed by a list of *key words and phrases* describing the subject matter of the article and taken from it. A list of the numbers may be found in the annual index of *Mathematical Reviews*, published with the December issue starting in 1990, as well as from the electronic service e-MATH [telnet **e-MATH.ams.org** (or telnet **130.44.1.100**). Login and password are **e-math**]. For journal abbreviations used in bibliographies, see the list of serials in the latest *Mathematical Reviews* annual index. When the manuscript is submitted, authors should supply the editor with electronic addresses if available. These will be printed after the postal address at the end of each article.

Electronically prepared manuscripts. The AMS encourages submission of electronically prepared manuscripts in $\mathcal{A}_{\mathcal{M}}\mathcal{S}$-TEX or $\mathcal{A}_{\mathcal{M}}\mathcal{S}$-LATEX because properly prepared electronic manuscripts save the author proofreading time and move more quickly through the production process. To this end, the Society has prepared "preprint" style files, specifically the amsppt style of $\mathcal{A}_{\mathcal{M}}\mathcal{S}$-TEX and the amsart style of $\mathcal{A}_{\mathcal{M}}\mathcal{S}$-LATEX, which will simplify the work of authors and of the production staff. Those authors who make use of these style files from the beginning of the writing process will further reduce their own effort. Electronically submitted manuscripts prepared in plain TeX or LaTeX do not mesh properly with the AMS production systems and cannot, therefore, realize the same kind of expedited processing. Users of plain TeX should have little difficulty learning $\mathcal{A}_{\mathcal{M}}\mathcal{S}$-TeX, and LaTeX users will find that $\mathcal{A}_{\mathcal{M}}\mathcal{S}$-LaTeX is the same as LaTeX with additional commands to simplify the typesetting of mathematics.

Guidelines for Preparing Electronic Manuscripts provides additional assistance and is available for use with either $\mathcal{A}_{\mathcal{M}}\mathcal{S}$-TeX or $\mathcal{A}_{\mathcal{M}}\mathcal{S}$-LaTeX. Authors with FTP access may obtain *Guidelines* from the Society's Internet node e-MATH.ams.org (130.44.1.100). For those without FTP access *Guidelines* can be obtained free of charge from the e-mail address guide-elec@ math.ams.org (Internet) or from the Customer Services Department, American Mathematical Society, P.O. Box 6248, Providence, RI 02940-6248. When requesting *Guidelines*, please specify which version you want.

At the time of submission, authors should indicate if the paper has been prepared using $\mathcal{A}_{\mathcal{M}}\mathcal{S}$-TeX or $\mathcal{A}_{\mathcal{M}}\mathcal{S}$-LaTeX. The *Manual for Authors of Mathematical Papers* should be consulted for symbols and style conventions. The *Manual* may be obtained free of charge from the e-mail address cust-serv@math.ams.org or from the Customer Services Department, American Mathematical Society, P.O. Box 6248, Providence, RI 02940-6248. The Providence office should be supplied with a manuscript that corresponds to the electronic file being submitted.

Electronic manuscripts should be sent to the Providence office immediately after the paper has been accepted for publication. They can be sent via e-mail to pub-submit@math.ams.org (Internet) or on diskettes to the Publications Department, American Mathematical Society, P. O. Box 6248, Providence, RI 02940-6248. When submitting electronic manuscripts please be sure to include a message indicating in which publication the paper has been accepted.

Two copies of the paper should be sent directly to the appropriate Editor and the author should keep one copy. The *Guide for Authors of Memoirs* gives detailed information on preparing papers for *Memoirs* and may be obtained free of charge from the Editorial Department, American Mathematical Society, P. O. Box 6248, Providence, RI 02940-6248. For papers not prepared electronically, model paper may also be obtained free of charge from the Editorial Department.

Any inquiries concerning a paper that has been accepted for publication should be sent directly to the Editorial Department, American Mathematical Society, P. O. Box 6248, Providence, RI 02940-6248.

Editors

This journal is designed particularly for long research papers (and groups of cognate papers) in pure and applied mathematics. Papers intended for publication in the *Memoirs* should be addressed to one of the following editors:

Ordinary differential equations, partial differential equations, and applied mathematics to JOHN MALLET-PARET, Division of Applied Mathematics, Brown University, Providence, RI 02912-9000; e-mail: am438000@brownvm.brown.edu.

Harmonic analysis, representation theory, and Lie theory to ROBERT J. STANTON, Department of Mathematics, The Ohio State University, 231 West 18th Avenue, Columbus, OH 43210-1174; e-mail: stanton@math.ohio-state.edu.

Ergodic theory, dynamical systems, and abstract analysis to DANIEL J. RUDOLPH, Department of Mathematics, University of Maryland, College Park, MD 20742; e-mail: djr@math.umd.edu.

Real and harmonic analysis to DAVID JERISON, Department of Mathematics, MIT, Rm 2–180, Cambridge, MA 02139; e-mail: jerison@math.mit.edu.

Algebra and algebraic geometry to EFIM ZELMANOV, Department of Mathematics, University of Wisconsin, 480 Lincoln Drive, Madison, WI 53706-1388; e-mail: zelmanov@math.wisc.edu

Algebraic topology and differential topology to MARK MAHOWALD, Department of Mathematics, Northwestern University, 2033 Sheridan Road, Evanston, IL 60208-2730; e-mail: mark@math.nwu.edu.

Global analysis and differential geometry to ROBERT L. BRYANT, Department of Mathematics, Duke University, Durham, NC 27706-7706; e-mail: bryant@math.duke.edu.

Probability and statistics to RICHARD DURRETT, Department of Mathematics, Cornell University, White Hall, Ithaca, NY 14853-7901; e-mail: rtd@cornella.cit.cornell.edu.

Combinatorics and Lie theory to PHILIP J. HANLON, Department of Mathematics, University of Michigan, Ann Arbor, MI 48109-1003; e-mail: phil.hanlon@math.lsa.umich.edu.

Logic and universal algebra to GREGORY L. CHERLIN, Department of Mathematics, Rutgers University, Hill Center, Busch Campus, New Brunswick, NJ 08903; e-mail: cherlin@math.rutgers.edu.

Algebraic number theory, analytic number theory, and automorphic forms to WEN-CHING WINNIE LI, Department of Mathematics, Pennsylvania State University, University Park, PA 16802-6401.

Complex analysis and nonlinear partial differential equations to SUN-YUNG A. CHANG. Department of Mathematics, University of California at Los Angeles, Los Angeles, CA 90024-1555; e-mail: chang@math.ucla.edu.

All other communications to the editors should be addressed to the Managing Editor, PETER SHALEN, Department of Mathematics, Statistics, and Computer Science, University of Illinois at Chicago, Chicago, IL 60680; e-mail: U10123@uicvm.uic.edu.

Recent Titles in This Series

(Continued from the front of the publication)

(See the AMS catalog for earlier titles)

Forsyth Library
Fort Hays State University